Lecture Notes of the Institute for Computer Sciences, Social Informatics and Telecommunications Engineering 185

Editorial Board

Ozgur Akan
  Middle East Technical University, Ankara, Turkey
Paolo Bellavista
  University of Bologna, Bologna, Italy
Jiannong Cao
  Hong Kong Polytechnic University, Hong Kong, Hong Kong
Geoffrey Coulson
  Lancaster University, Lancaster, UK
Falko Dressler
  University of Erlangen, Erlangen, Germany
Domenico Ferrari
  Università Cattolica Piacenza, Piacenza, Italy
Mario Gerla
  UCLA, Los Angeles, USA
Hisashi Kobayashi
  Princeton University, Princeton, USA
Sergio Palazzo
  University of Catania, Catania, Italy
Sartaj Sahni
  University of Florida, Florida, USA
Xuemin Sherman Shen
  University of Waterloo, Waterloo, Canada
Mircea Stan
  University of Virginia, Charlottesville, USA
Jia Xiaohua
  City University of Hong Kong, Kowloon, Hong Kong
Albert Y. Zomaya
  University of Sydney, Sydney, Australia

T0212475

More information about this series at http://www.springer.com/series/8197

Joaquim Ferreira · Muhammad Alam (Eds.)

Future Intelligent Vehicular Technologies

First International Conference, Future 5V 2016
Porto, Portugal, September 15, 2016
Revised Selected Papers

 Springer

Editors
Joaquim Ferreira
University of Aveiro
Campus Universitário de Santiago
Aveiro
Portugal

Muhammad Alam
Instituto de Telecomunicações
Campus Universitário de Santiago
Aveiro
Portugal

ISSN 1867-8211   ISSN 1867-822X (electronic)
Lecture Notes of the Institute for Computer Sciences, Social Informatics
and Telecommunications Engineering
ISBN 978-3-319-51206-8   ISBN 978-3-319-51207-5   (eBook)
DOI 10.1007/978-3-319-51207-5

Library of Congress Control Number: 2016962018

© ICST Institute for Computer Sciences, Social Informatics and Telecommunications Engineering 2017
This work is subject to copyright. All rights are reserved by the Publisher, whether the whole or part of the
material is concerned, specifically the rights of translation, reprinting, reuse of illustrations, recitation,
broadcasting, reproduction on microfilms or in any other physical way, and transmission or information
storage and retrieval, electronic adaptation, computer software, or by similar or dissimilar methodology now
known or hereafter developed.
The use of general descriptive names, registered names, trademarks, service marks, etc. in this publication
does not imply, even in the absence of a specific statement, that such names are exempt from the relevant
protective laws and regulations and therefore free for general use.
The publisher, the authors and the editors are safe to assume that the advice and information in this book are
believed to be true and accurate at the date of publication. Neither the publisher nor the authors or the editors
give a warranty, express or implied, with respect to the material contained herein or for any errors or
omissions that may have been made. The publisher remains neutral with regard to jurisdictional claims in
published maps and institutional affiliations.

Printed on acid-free paper

This Springer imprint is published by Springer Nature
The registered company is Springer International Publishing AG
The registered company address is: Gewerbestrasse 11, 6330 Cham, Switzerland

Preface

The incorporation of information and communication technologies within vehicles and transportation infrastructure will revolutionize the way we travel. The enabling technologies are intended to realize the frameworks that will spur an array of applications and use cases in the domain of road safety, traffic efficiency, and driver's assistance. These applications will allow for the dissemination and gathering of useful information among vehicles and between transportation infrastructure and vehicles in pursuance of assisting people to travel safely and comfortably. Although transportation systems are evolving toward intelligent transportation systems, they face critical challenges and need to be addressed to emerge as intelligent vehicular technologies. Therefore, the European Alliance for Innovation (EAI) has taken a step toward the realization of future intelligent vehicular technologies by hosting both the academic and the industrial research communities at the Future 5V conference in Porto, Portugal. Future 5V is an annual international conference organized by the EAI (European Alliance for Innovation) and co-sponsored by Springer. The central theme of the conference is focused on sharing with the research community new paradigms and proposals for future intelligent vehicular technologies, which are considered to be the key research area in the intelligent transportation systems. Future 5V welcomes research articles in the field of vehicular networks/communications covering theory and practice in the aforementioned field of study.

Future 5V 2016 hosted the "Internet of Things (IoT) Meets Big Data and Cloud Computing (IoT-BC)" workshop, which further extended the domain of the conference and attracted more researchers worldwide. The IoT is the next wave in the era of computing outside the realm of traditional desktop. The IoT ecosystem includes any form of technology that can connect to the Internet. This means connected cars, wearables, TVs, smartphones, fitness equipment, robots, ATMs, vending machines, and all of the vertical applications, security and professional services, analytics, and platforms that come with them. The presence of embedded sensor nodes and the assignment of IP addresses make these physical objects smart enough to interact and share data. Considering the fact that millions or perhaps billions of such objects will be connected with the Internet, the volume of data generated will be enormous. The data generated on an unprecedented scale may face peculiar security issues, which may not be possible to be addressed with the existing security mechanisms. The data may be highly redundant and may require highly efficient analysis tools to extract the useful data. The existing tools for data analysis and extraction may not suffice this requirement and therefore tools such as Hadoop and sensor-fitted devices are going to receive a lot of attention. The data generated by the sheer number of devices in an IoT paradigm would require abundant memory storage. However, the devices are sensor embedded and as such have restrictions on various resources such as computation, storage, available bandwidth and battery power. As a result, the data need to be stored on distributed clouds. The existing security schemes incorporated at various clouds may not be efficient for the data of real-world physical objects because the existing schemes dealing with traditional network data may not work with data coming from physical objects such as refrigerators, TVs, and fitness equipment. Hence, lightweight but efficient security algorithms need to be designed for handling data of such objects on distributed clouds.

The Future 5V 2016 Technical Program Committee comprised more than 30 leading experts in their field, and was assisted by 20 additional external reviewers originating from different countries worldwide. Beside the main track, the conference hosted an international workshop that further extended the scope of the conference and its exposure to international research community. Future 5V 2016 registered more than 30 attendees and attracted more than 50 paper submissions that were peer reviewed by independent experts. Professor Jaime Lloret served as the conference keynote speaker and delivered a talk on "Artificial Intelligence in Vehicular Ad Hoc Networks." Beside the keynote talk, the conference hosted a panel discussion session on "Vehicular Communication and 5G Paradigm" in which more than 35 researchers participated and thoroughly discussed the existing vehicular communication standards, their potentials and shortcomings, and the roadmap toward future vehicular communication that is envisioned in the 5G paradigm.

November 2016   Joaquim Ferreira
Muhammad Alam

Conference Organization

Conference General Chair

Joaquim Ferreira   University of Aveiro, Aveiro, Portugal

Conference General Co-chairs

Muhammad Alam   Instituto de Telecomunicações, Aveiro, Portugal
Elad Schiller   Chalmers University of Technology, Sweden

Steering Committee

Imrich Chlamtac   Create-Net, Italy
Muhammad Alam   Instituto de Telecomunicações, Aveiro, Portugal

Publicity Chair

Mithun Mukherjee   Guangdong University of Petrochemical Technology,
                                    China

Publicity Co-chair

Chunsheng Zhu   University of British Columbia, Canada

Social Media Chair

Bruno Silva   Instituto de Telecomunicações, Aveiro, Portugal

Technical Program Committee Chairs

Nadir Shah   COMSATS Institute of Information Technology,
                                    Pakistan
Paulo Pedreiras   University of Aveiro, Portugal
Wael Dghais   Higher Institute of Applied Sciences and Technology
                                    of Sousse, Tunisia

Web Chairs

Bilal Habib   George Mason University, USA
Awais Jadoon   Instituto de Telecomunicações, Aveiro, Portugal

Workshops Chairs

Luis Almeida	University of Porto, Portugal
Giovanni Iovino	INTECS, Italy
Luis Silva	Instituto de Telecomunicações, Aveiro, Portugal

Poster Chair

Yuanfang Chen	Guangdong University of Petrochemical Technology, China

Demos Chair

Xiaoling Wu	Guangzhou Institute of Advanced Technology, Chinese Academy of Sciences, China

Panels Chair

Paulo Pedreiras	University of Aveiro, Aveiro, Portugal

Sponsorship and Exhibits Chair

João Almeida	University of Aveiro, Portugal

Publication Chair

Faisal Bashir	Bahria University, Islamabad, Pakistan
Muhammad Alam	Instituto de Telecomunicações, Aveiro, Portugal

Local Chair

Mushtaq Raza	University of Porto, Portugal
Muhammad Ali Khan	University of Porto, Portugal
Niaz Ali Khan	University of Porto, Portugal

Local Host

Instituto de Telecomunicações - Embedded Systems, Aveiro, Portugal

Panel Session Chair

José Fonseca	University of Aveiro, Aveiro, Portugal

Conference Manager

Lenka Laukova	EAI - European Alliance for Innovation

Technical Program Committee Chair

Nadir Shah   COMSATS Institute of Information Technology,
                                    Pakistan

Technical Program Committee Co-chairs

Wael Dghais   Higher Institute of Applied Sciences and Technology
                                    of Sousse, Tunisia
Paulo Pedreiras   University of Aveiro, Aveiro, Portugal

Technical Program Committee Members

M. Imran Hayee	University of Minnesota Duluth, USA
Mohammad Hossein Anisi	University of Malaya, Malaysia
Junaid Arshad	University of West London, UK
Paolo Bellavista	DISI - University of Bologna, Italy
Xiaoling Wu	Guangzhou Institute of Advanced Technology, Chinese Academy of Sciences, China
Shahid Mumtaz	Instituto de Telecomunicações, Portugal
Bilal Habib	George Mason University, USA
Mian Ahmad	UTS, Australia
Kazi Huq	Instituto de Telecomunicações, Aveiro, Portugal
Yizhen Liu	U.S. Research Institute of Huawei at Santa Clara, California, USA
Yuanfang Chen	Guangdong University of Petrochemical Technology, China
Fazlullah Khan	Abdul Wali Khan University, KPK, Pakistan
Mazliza Othman	University of Malaya, Malaysia
Mushtaq Raza	University of Porto, Portugal
Mubashir Husain Rehmani	COMSATS Institute of Information Technology, Wah, Pakistan
Luis Almeida	University of Porto, Portugal
Atta ur Rehman Khan	King Saud University, Saudi Arabia
Elad Schiller	Chalmers University of Technology, Sweden
Muhammad Ajmal Azad	INESC TEC, FEUP, Porto, Portugal
Chunsheng Zhu	University of British Columbia, Canada
Nadia Nawaz Qadri	COMSATS Institute of Information Technology, Wah, Pakistan
Saeed Ullah	Kyung Hee University, South Korea
Muhammad Alam	Instituto de Telecomunicações, Aveiro, Portugal
Mukhtaj Khan	Brunel University, UK
Ihsan Ali	University of Malaya, Malaysia
Izaz ur Rahman	Brunel University, UK
Bruno Silva	Instituto de Telecomunicações, Aveiro, Portugal

Nadir Shah	COMSATS Institute of Information Technology, Pakistan
Paulo Pedreiras	University of Aveiro, Aveiro, Portugal
Wael Dghais	Higher Institute of Applied Sciences and Technology of Sousse, Tunisia
Syed Aftab Rashid	CISTER Research Center, Porto, Portugal
Mithun Mukherjee	Guangdong University of Petrochemical Technology, China
Awais Jadoon	Instituto de Telecomunicações, Aveiro, Portugal
Giovanni Iovino	INTECS, Italy
Luis Silva	Instituto de Telecomunicações, Aveiro, Portugal
Mohammad Saiful Islam Mamun	University of New Brunswick, Canada
Shahbaz Akhtar Abid	COMSATS Institute of Information Technology, Lahore, Pakistan
João Almeida	University of Aveiro, Portugal
Byung-Seo Kim	Hongik University, South Korea
Ismail Ahmedy	University of Malaya, Malaysia

Workshop: Internet of Things (IoT) Meets Big Data and Cloud Computing

General Chair

Ihsan Ali	Abdul Wali Khan University, Mardan, Pakistan
Fazlullah Khan	Abdul Wali Khan University, Mardan, Pakistan

General Co-chair

Mian Ahmad Jan	Abdul Wali Khan University, Mardan, Pakistan

Session Chairs

Mukhtaj Khan	Abdul Wali Khan University, Mardan, Pakistan
Izaz ur Rahman	Abdul Wali Khan University, Mardan, Pakistan

Technical Program Committee Chair

Fazlullah Khan	Abdul Wali Khan University, Mardan, Pakistan

Technical Program Committee Co-chair

Mian Ahmad Jan	University of Technology Sydney, Australia

Technical Program Committee Members

Fazlullah Khan	Abdul Wali Khan University Mardan, Pakistan
Mian Ahmad Jan	Abdul Wali Khan University Mardan, Pakistan
Mukhtaj Khan	Abdul Wali Khan University Mardan, Pakistan
Izaz ur Rahman	Abdul Wali Khan University Mardan, Pakistan

Muhammad Usman Khan	University of Technology Sydney, Australia
Mahardhika Pratina	LaTrobe University, Australia
Xhiyan (Thomas) Tan	University of Twente, The Netherlands
Thawatchai Chomsiri	Mahasangam Univerdity, Thailand
Mohammed Ambu Sadi	College of Applied Science, Nizwa, Oman
Khaled Al-Debei	University of Jordan
Adrian Johannas	University of PGRI Ronggolawe Tuban, Indonesia
Aaiza Gul	University of Technology, Malaysia
Attiq ur Rahman	Southampton University, UK
Muhammad Zakarya	University of Surrey, UK
Muhammad Khan	Brunel University, UK
Nabeel Younas Khan	University of Auckland, New Zealand
Muhammad Al-Shehri	Taif University, Kingdom of Saudi Arabia
Deepak Puthan	CSIRO, Australia
Chi Yang	CSIRO, Australia
Asfandiyar Khan	Tomsk University, Russia
Forhad ud Din	University of Dhaka, Bangladesh

Contents

Workshop on Internet of Things (IoT) meets Big Data and Cloud Computing

Intelligent Vehicular Communication

Enforcing Replica Determinism in the Road Side Units of Fault-Tolerant Vehicular Networks

João Almeida[1,2]([✉]), Joaquim Ferreira[1,3], Arnaldo S.R. Oliveira[1,2],
Paulo Pedreiras[1,2], and José Fonseca[1,2]

[1] Instituto de Telecomunicações, Aveiro, Portugal
[2] DETI, Universidade de Aveiro, Aveiro, Portugal
[3] ESTGA, Universidade de Aveiro, Águeda, Portugal
{jmpa,jjcf,arnaldo.oliveira,pbrp,jaf}@ua.pt

Abstract. This paper presents a strategy to enforce replica determinism in the road-side units (RSUs) of wireless vehicular networks. An active replication scheme is used to enhance fault-tolerant behaviour in these RSU nodes, which are responsible for handling channel access and admission control policies in real-time vehicular communications protocols. The proposed solution guarantees consistency among all RSU replicas, by introducing a dedicated link shared exclusively by these units, allowing them to implement an atomic commit protocol of the packets received through the wireless medium. This strategy also has the advantage of reducing packet loss, since only one replica needs to successfully decode the packet in order for it to become available to the RSU group of replicas. It should be noticed however that this method increases the total delay of packet delivery to the upper layers of the communications protocol, so its impact on the real-time properties of the network needs to be further evaluated.

Keywords: Vehicular networks · Intelligent transportation systems · Real-time communications · Fault-tolerance mechanisms · Active replication scheme · Replica consistency · Atomic commit protocol

1 Introduction

Cooperative Intelligent Transportation Systems (C-ITS) aim to reduce road accidents, by extending vehicle's field of view and producing warnings alerts in case of dangerous traffic situations. In addition, these systems have the goal of decreasing CO_2 emissions, road congestions and energy consumption. Infotainment applications based on C-ITS can also be developed, in order to improve passenger's comfort and to provide a better riding and driving experience. Vehicular communications are the main enabling technology supporting these collaborative systems, since they allow vehicles to communicate among each others, to exchange information with the road-side infrastructure and eventually with pedestrians, cyclists and other objects located close to the roads. This is the

© ICST Institute for Computer Sciences, Social Informatics and Telecommunications Engineering 2017
J. Ferreira and M. Alam (Eds.): Future 5V 2016, LNICST 185, pp. 3–12, 2017.
DOI: 10.1007/978-3-319-51207-5_1

essential concept behind Internet of Vehicles (IoV) [8], which constitutes a sub-group of the future Internet of Things (IoT). In the IoV, at least a significant percentage of vehicles will be equipped with these communications capabilities, and will be able for instance to disseminate an accident detection, avoiding further chain collisions. Another advantage is that vehicles will be connected with other networks, for example with the sensors network of the driver's home, allowing him/her to control the heat or air conditioning systems, the garage door, the lightning and surveillance systems, etc. before arriving home [7].

Vehicular networks are based on the IEEE WAVE protocol stack in USA and on the ETSI ITS-G5 in Europe. Both of them rely on the IEEE 802.11 standard [9] for the physical and medium access control (MAC) layers, the same standard employed in Wi-Fi technology, but with some different settings, namely the reduced channel bandwidth (10 MHz instead of 20 MHz) to mitigate the impacts of multipath and Doppler effects, and the absence of authentication and association procedures for faster link establishment due to the short connections in these very dynamic environments. The use of the Carrier Sense Multiple Access with Collision Avoidance (CSMA/CA) method, defined in the standard for managing channel access, resulted in several studies [6, 10] reporting multiple problems associated with this mechanism under congested traffic scenarios. Some of these issues are related with the high number of packet collisions and the large values for the end-to-end delay. In order to deal with this problem, ETSI for instance has proposed a Decentralized Congestion Control mechanism that still relies on the CSMA/CA scheme but enables a fine control of some variables that directly influence channel occupation, such as receiver sensitivity, transmit power level, etc. Other works in the literature [5, 11] have proposed deterministic MAC protocols to support real-time wireless communications in these safety-critical environments. Due to the higher reliability they provide, solutions based on the support provided by the road-side infrastructure are typically more suitable for vehicular scenarios, where the network's topology changes very quickly. The reference nodes placed in fixed locations, named road-side units (RSUs), can handle admission control policies and the whole traffic scheduling from the on-board units (OBUs) placed inside the vehicles, as suggested in [11]. However, since in this scheme, the RSUs control all network communications, these nodes become critically important for the operation of the intelligent traffic system. As a result, a fault-tolerant vehicular network has been proposed [3], essentially targeting the role played by these master nodes and aiming to increase dependability of the overall network. An active replication scheme for the RSUs was deployed [2], enabling fast recovery of the RSU failed node, by using a backup replica to replace the operation of the primary unit. Nevertheless, in order to work properly, this mechanism assumes that both the primary and the backup replicas are synchronized with respect to the information both units hold about the network. For that to be true, all replicas need to receive and process the same packets send by all other vehicular nodes. Since the wireless medium is not totally reliable, even if the replicas are co-located, they may not be able to decode the same packets without errors. Consequently, this works focus on guaranteeing replica consistency by employing an atomic commit protocol to disseminate all the messages received by both RSU replicas.

2   Dependable Wireless Vehicular Networks

This section provides the background of this work, describing the operation of a real-time vehicular communications protocol for which a fault-tolerant infrastructure-based architecture was designed, aiming to improve the dependability attributes of the network. A detailed description of the proposed RSU replication scheme is provided, for which the replica consistency strategy presented in this paper was developed.

2.1   Real-Time Vehicular Communications Protocol

A deterministic MAC protocol for infrastructure-based vehicular networks was proposed in [11]. This protocol, named Vehicular Flexible-Time-Triggered (V-FTT), is based on a spatial TDMA scheme with a multi-master multi-slave architecture. A backhauling network interconnects the RSUs (master nodes), enabling them with a global vision of the road traffic situation. In V-FTT, time is divided in consecutive Elementary Cycles (ECs) with 50 ms duration, as depicted in Fig. 1. Each EC is further divided into three main windows. The first interval (Infrastructure Window) is used by RSUs to transmit warning messages concerning road hazards together with packets containing scheduling information to the OBUs (slave nodes). These scheduling messages include the time slot assignment for the transmission of OBUs' packets. After the Infrastructure Window, there is a congestion based interval named Free Period, during which vehicles can register with the road infrastructure and nodes non-compliant with V-FTT protocol can transmit. At the end of each EC (Synchronous OBU Window), the OBUs send safety messages according to the time slot scheduling previously disseminated. These packets include information about the vehicle's state (e.g. position, speed and orientation) and the perceived environment (e.g. dangerous traffic situations).

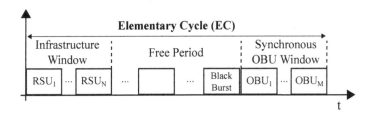

Fig. 1. V-FTT protocol.

2.2 Fault-Tolerant Network Architecture

Based on the need to provide strict real-time and dependability requirements for safety-critical vehicular applications, a fault-tolerant infrastructure-based network architecture was proposed in [3]. Given the key role played by the RSUs in the admission control and scheduling processes, a fail silence mechanism was designed [1], ensuring that these units can only fail by not sending any message to the network. After guaranteeing fail silent behaviour, an active replication scheme was developed [2], providing low-delay recovery procedure of the failed RSU nodes. On the OBU side, a medium guardian entity was devised in order to constrain packet transmission solely for the period inside the time slot previously allocated for that specific mobile unit. A diagram of the fault-tolerant architecture comprising these three major mechanisms is depicted in Fig. 2.

2.3 Active RSU Replication Scheme

The active replication scheme proposed in [2] for the RSUs of infrastructure-based vehicular networks guarantees that in case of failure in the primary unit, a backup replica will replace its operation within a small time interval ($\approx 25~\mu$s). This way, there is no discontinuity of the traffic scheduling and the real-time guarantees provided by deterministic MAC protocols such as V-FTT, can still be delivered. The proposed mechanism works in the following way. The packet transmission of the backup RSU is always delayed by a small interval in relation to the primary transmission. In this sense, the backup RSU will sense the wireless medium and, if it is occupied, it knows that the primary replica is free of error. Otherwise, if the medium is free, the backup transmits the message, considering that the primary unit failed to send the message. This is only possible due to

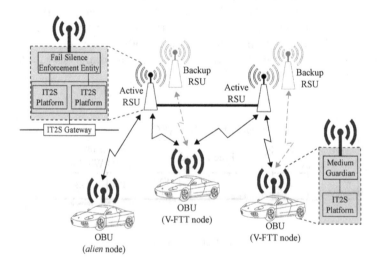

Fig. 2. Fault-tolerant vehicular network architecture [2].

the fail-silent behaviour of the RSU replicas, both in value and time domains. However, for the correct operation of this replication scheme, it is necessary that both replicas are synchronized in time, which is ensured by local GPS receivers, and with respect to the information they hold about the current road traffic situation, e.g. number and identification of vehicles in the coverage area. This last premise is not always valid, since due to the unpredictability of the wireless medium in such dynamic environments, the successfully received messages may not be the same in both replicas, even considering that they are co-located. Therefore, a new strategy to ensure replica consistency between the primary and backup units of each RSU node is proposed in this paper.

3   RSU Replica Determinism

The main goal behind ensuring replica determinism is to guarantee that in case of failure in one of the replicas, the remaining ones continue to deliver the same service without causing any inconsistency in the system. For that purpose, all the replicas need to be synchronized, sharing a consistent view of the environment in which they operate. In the case of infrastructure-based vehicular communications, the distinct RSU replicas must hold the same information about the current state of the road traffic network. For that purpose, they need to successfully decode the same packets and process these messages in a deterministic way. Otherwise, the vehicular network database residing in both replicas could differ and lead to inconsistent sequences of exchanged messages. Let's consider for instance, the situation in which a vehicle transmits a message to register with the network. Assume that this message is correctly received by the active RSU, but in the backup unit, there is an error in the decoding process. As a result, the active node will generate and send a new identification number to this OBU and will schedule a time slot for this new registered vehicle to transmit in the next Elementary Cycle. Then, imagine that the active RSU fails, being automatically replaced by the backup unit. In this scenario, the backup replica will not recognize the messages sent by this new OBU and will not include it in the future slot assignments, since it was not previously registered in the vehicular network database of the backup RSU. This situation is unacceptable and cannot be allowed in the operation of the real-time communications protocol.

In order to solve this problem, this work proposes the introduction of a dedicated wired link between the active and backup replicas, with the main goal of ensuring the correct dissemination of all messages successfully received by at least one of the replicas. This way, it is possible to implement a distributed atomic commitment of the packets received through the wireless medium without utilizing additional network resources (e.g. channel bandwidth). Figure 3 shows the proposed solution, in which correctly decoded packets are delivered to both replicas by employing an atomic commit protocol.

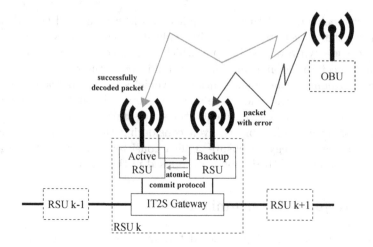

Fig. 3. Replica consistency strategy in infrastructure-based vehicular networks.

3.1 Distributed Atomic Commitment

In distributed systems, an atomic commitment is traditionally achieved by utilizing the two-phase or the three-phase commit protocols or some variant of these. The two-phase commit (2PC) protocol [4, Chap. 7] consists of two communication phases that are coordinated by one of the participants. Initially, during the prepare or voting phase, the coordinator sends a VOTE-REQ (i.e., vote request) message to all other participants. Upon reception of a VOTE-REQ message, the participant replies with a YES or NO message, depending if it is ready or not to commit the transaction. If the answer is negative, the participant immediately aborts the transaction. In the second phase, also known as commit phase, if the coordinator has received at least one negative reply, it will abort the transaction and send an ABORT message. On the other hand, if the coordinator has only received affirmative responses from all the participants, it will commit the transaction and will send a COMMIT message. In both cases (COMMIT or ABORT), the participants will respond with an acknowledgement message after having acted accordingly to the received command.

Despite its efficiency, the two-phase commit protocol has a main drawback. In case of failure of the coordinator during the transition period between the two phases, all the participants will remain blocked waiting for the decision. In order to solve this problem, the three-phase commit (3PC) protocol [12] was proposed, providing a non-blocking solution. In the 3PC protocol, an additional intermediate phase is introduced between the prepare and commit steps of the two-phase commit protocol. This extra phase is usually designated as pre-commit and has the main objective of delaying the final decision until all currently active participants are aware of which resolution the coordinator is about to take. In this scheme, the coordinator sends a PRE-COMMIT message to all participants if the prepare phase is successful. Then, after another round of acknowledgements, the coordinator finally moves to the commit phase. In case of a missing acknowledgement message, the coordinator will abort the transaction. On the other hand, if a participant does not receive the final COMMIT message after a successful pre-commit phase, it will automatically commit the transaction. This way, the blocking situation that can occur in the two-phase commit protocol will be avoided by introducing this intermediate phase.

3.2   Atomic Commit Protocol

The distributed atomic commitment protocol proposed in this paper is inspired in the previously described solutions. However, given the specific characteristics of the scenario presented in Fig. 3, some aspects of the proposed protocol are different from the traditional ones. For instance, in the developed RSU replication scheme there are only two replicas, which means that in each transaction (corresponding to a packet received through the wireless interface), besides one coordinator, there is only one participant in the atomic commit protocol. Additionally, the scenario depicted in Fig. 3 poses an additional challenge, giving the fact that there are two interfaces by which the RSU replicas can communicate, i.e. the wireless and the cabled one. Consequently, several situations may occur, since at a given instant both interfaces can be working perfectly, both can have failed or one can be running while the other is down. Moreover and in order to comply with the real-time requirements of vehicular environments, the implemented replication strategy follows a low latency recovery procedure, which leads to the presence of an active and a backup replica with distinct functional behaviours. For example, since the active RSU is the only replica to transmit messages over the wireless medium, failures in the wireless communications interface of the backup unit can not be detected by the active one.

Given these circumstances, a careful protocol design must be followed in order to ensure a reliable and consistent operation of the road-side infrastructure under

---

Algorithm 1. Active RSU's algorithm

---

if *packet received through the wireless interface* **then**
    # 1 → *commit packet information*
    # 2 → *send packet to the backup RSU*
    # 3 → *wait for an acknowledgment message from the backup replica*
    if *timeout expired before acknowledgment message received* **then**
        # 4 → *stop sending any type of messages to the backup RSU*
    end if
else if *packet received through the dedicated wired link* **then**
    # 1 → *commit packet information*
    # 2 → *send acknowledgement message to the backup RSU*
else if *no packet sent to the backup RSU in the last EC period* **then**
    # 1 → *send keepAlive message*
else if *no packet received from the backup RSU in the last EC period* **then**
    # 1 → *inform upper layers about the faulty link/replica*
end if

Algorithm 2. Backup RSU's algorithm

---

if *packet received through the wireless interface* **then**
    # 1 → *send packet to the active RSU*
    # 2 → *wait for an acknowledgment message from the active replica*
    if *acknowledgment message received* **then**
        # 3 → *commit packet information*
    else if *timeout expired before acknowledgment message received* **then**
        # 3 → *inform upper layers about the faulty link/replica*
        if *active RSU is still transmitting in the wireless interface* **then**
            # 4 → *stop backup replica operation*
        else
            # 4 → *commit packet information*
        end if
    end if
else if *packet received through the dedicated wired link* **then**
    # 1 → *commit packet information*
    # 2 → *send acknowledgement message to the active RSU*
else if *no packet sent to the active RSU in the last EC period* **then**
    # 1 → *send keepAlive message*
else if *no packet received from the active RSU in the last EC period* **then**
    # 1 → *inform upper layers about the faulty link/replica*
    if *active RSU is still transmitting in the wireless interface* **then**
        # 2 → *stop backup replica operation*
    end if
end if

---

all different fault scenarios. To simplify the design process, it is assumed that each RSU replica presents fail-silent behaviour not only in the wireless interface [1], but also in the dedicated wired link. This condition is not guaranteed by the previously developed RSU architecture, however it can be easily achieved by implementing a fail silence enforcement entity, similar to the one devised in [1], for the output traffic of the replica's cabled connection. This means that only valid packets will be transmitted within specified time intervals. In addition, it is also assumed that all packets correctly delivered by the atomic commit protocol are posteriorly handled in an upper layer of the protocol stack, before the information contained in the received messages is included in the RSU replica's database. In this step, it is verified for instance if there is a duplication of the packets received both from the wireless interface and the dedicated link between the replicas. Under perfect circumstances, all the packets would be received through both interfaces, so it is necessary to discard the duplicated messages.

The operation of the proposed atomic commit protocol, both in the active and in the backup replicas, is depicted by the Algorithms 1 and 2, respectively. In the absence of faults, all the packets received through the wireless interfaces of both replicas will be first committed in the active RSU, in order to avoid any inconsistency in case of failure of the active replica. Therefore, it should be noticed that if the active RSU stops working, the backup replica will only replace its operation, if it has delivered the same packets to the upper layers of the protocol stack. This is the reason why in Algorithm 1, all the packets received from both the wireless interface and the dedicated link in the active RSU, are immediately committed. Then, if the packet was received through the wireless interface, it will be sent to the backup RSU using the dedicated wired link. The active RSU will await for an acknowledgement message, that if not arrives within a maximum time limit (and after some retransmissions), will cause the active replica to completely stop sending messages to the backup unit. As a result, this is will force the backup RSU to stop its operation (Algorithm 2). In the other the hand, if the packet was received through the wired link, the active replica will just reply with an acknowledgement. Moreover and in order to verify the connectivity status between the replicas when there are no messages transmitted in the wireless medium, the active replica will transmit a *keepAlive* message at least once per Elementary Cycle (EC). On the other hand, if no packet is received from the backup unit during an entire EC, the active RSU will inform the upper layers that there is a problem with the wired link or backup replica, information that will be forwarded to the backhauling network (via IT2S Gateway).

In the backup RSU (Algorithm 2), if a packet is received through the wireless interface, it will be first sent to the active replica, and it will only be committed after receiving the acknowledgement message. If this acknowledgement does not reach the backup unit, the information about the faulty link or replica will be forwarded to the upper layers. In this case, if the active RSU is still transmitting in the wireless interface, the backup replica will stop its operation, since the active unit is free of faults. However, if the wireless interface of the active RSU is silent, that means the backup is the only replica operating and therefore it can commit the packet information. When receiving a packet through the dedicated wired link, the backup RSU will simply deliver it to the upper layers and send an acknowledgement message to the active replica. The rest of the algorithm is similar to the one in the active RSU, except for the situation when no packets are received from the dedicated link during an entire EC. In this case, the backup replica will verify if the active unit is still transmitting in the wireless interface and it will stop working under those circumstances, in order to avoid any inconsistency in the event of a failure in the active RSU.

4   Conclusions and Future Work

In this paper, an atomic commit protocol is proposed in order to enforce replica determinism between the active and the backup units of an RSU node in infrastructure-based vehicular networks. The goal of this protocol is to disseminate the messages exchanged in the wireless medium through a dedicated wired link connecting both replicas. This way, all the packets received by at least one of the replicas will be delivered to the upper layers in both units. Given the distinct roles played by the active and backup replicas, the algorithms running in both units for the implementation of the atomic commit protocol will be different, as explained in the paper. As future work, the protocol needs to be implemented in a practical vehicular communications system and the timing overhead introduced by the operation of the protocol needs to be evaluated, as well as its impact in the performance of the real-time vehicular network.

Acknowledgements. This work is funded by National Funds through FCT - Fundação para a Ciência e a Tecnologia under the PhD scholarship ref. SFRH/BD/52591/2014 and the project PEst-OE/EEI/LA0008/2013.

References

1. Almeida, J., Ferreira, J., Oliveira, A.: Fail silence mechanism for dependable vehicular communications. Int. J. High Perform. Comput. Networking (2017) (in press)
2. Almeida, J., Ferreira, J., Oliveira, A.S.R.: An RSU replication scheme for dependable wireless vehicular networks. In: 12th European Dependable Computing Conference (EDCC 2016), Gothenburg, Sweden, 5-9 September 2016, pp. 229–240 (2016). doi:10.1109/EDCC.2016.11
3. Almeida, J., Ferreira, J., Oliveira, A.S.R.: Fault tolerant architecture for infrastructure based vehicular networks. In: Alam, M., Ferreira, J., Fonseca, J. (eds.) Intelligent Transportation Systems. SSDC, vol. 52, Chap. 8, pp. 169–194. Springer, Cham (2016)
4. Bernstein, P.A., Hadzilacos, V., Goodman, N.: Concurrency Control, Recovery in Database Systems. Addison-Wesley Longman Publishing Co., Inc., Boston, MA, USA (1987). ISBN: 0-201-10715-5
5. Böhm, A., Jonsson, M.: Real-time communication support for cooperative, infrastructure-based traffic safety applications. Int. J. Veh. Technol. **2011**, 1–17 (2011)
6. Bilstrup, K., et al.: On the ability of the 802.11p MAC method, STDMA to support real-time vehicle-to-vehicle communication. EURASIP J. Wireless Commun. Networking **2009**(1), 1–13 (2009)
7. BMW Group. BMW ConnectedDrive at the IFA 2015 consumer electronics show in Berlin. https://www.press.bmwgroup.com/global/article/detail/T0232769EN/bmwconnecteddrive-at-the-ifa-2015-consumer-electronics-show-inberlin. Accessed 28 May 2016
8. Gerla, M., et al.: Internet of vehicles: from intelligent grid to autonomous cars and vehicular clouds. In: IEEE World Forum on Internet of Things (WF-IoT), pp. 241–246 (2014)
9. IEEE Standard for Information Technology - Telecommunications, information exchange between systems Local, metropolitan area networks - Specific requirements Part 11: Wireless LAN Medium Access Control (MAC) and Physical Layer (PHY) Specifications. In: IEEE Std 802.11-2012 (Revision of IEEE Std 802.11-2007), pp. 1–2793 (2012)
10. Kloiber, B., et al.: Performance of CAM based safety applications using ITS-G5A MAC in high dense scenarios. In: Intelligent Vehicles Symposium (IV), pp. 654–660. IEEE (2011)
11. Meireles, T., Fonseca, J., Ferreira, J.: The case for wireless vehicular communications supported by roadside infrastructure. In: Perallos, A., et al. (eds.) Intelligent Transportation Systems Technologies and Applications, pp. 57–82 (2015)
12. Skeen, D., Stonebraker, M.: A formal model of crash recovery in a distributed system. IEEE Trans. Softw. Eng. **SE–9**(3), 219–228 (1983)

A Deterministic MAC Protocol
for Infrastructure to Vehicle Communications
in Motorways

Tiago Meireles[1(✉)], Joaquim Ferreira[2], and José Fonseca[3]

[1] Universidade da Madeira, Funchal, Portugal
hipkin@uma.pt
[2] Instituto de Telecomunicações/ESTGA,
Universidade de Aveiro, Aveiro, Portugal
jjcf@ua.pt
[3] Instituto de Telecomunicações/DETI,
Universidade de Aveiro, Aveiro, Portugal
jaf@ua.pt

Abstract. The current wireless standards devised for vehicle communications are not designed for hard real-time restrictions. In certain scenarios, such as motorways where a high density of vehicles travelling at high speed is common, the Medium Access Layer (MAC) of the existing standards is not deterministic and does not guarantee an upper bound for the delay of communications. This article discusses several proposals to address this issue and presents a deterministic MAC protocol for infrastructure to vehicle communications: the vehicular flexible time triggered protocol.

Keywords: Wireless vehicular communications · Real-time · Safety critical

1 Introduction

Recent developments in wireless communications devised for Intelligent Transportation Systems (ITS) makes possible to implement cooperative applications that can improve passenger's safety and comfort as well as traffic management. These networks rely on every vehicle having an on-board unit (OBU) capable of communication with other vehicles or with some kind of infrastructure on the road-side, also known as road-side unit (RSU). Vehicular safety applications have specific characteristics such as small latencies, as an example the Emergency Electronic Brake Light safety application requires that maximum latencies are lower than 100 ms. On the other hand, many multimedia applications require data rates higher than 1 Mbps and QoS support. The IEEE802.11p standard (which was included in 2012 in the amendment 6 of IEEE802.11 [1]), along with the IEEE 1609 set of standards was devised to respond to both the latency and throughput requirements of vehicle applications. Its European equivalent standard ETSI ITS G5 shares the same goal [2]. Their medium access control (MAC) layer adopts a carrier sense multiple access with collision avoidance (CSMA/CA), same as IEEE 802.11a, but with a new additional, non-IP, communication

© ICST Institute for Computer Sciences, Social Informatics and Telecommunications Engineering 2017
J. Ferreira and M. Alam (Eds.): Future 5V 2016, LNICST 185, pp. 13–23, 2017.
DOI: 10.1007/978-3-319-51207-5_2

protocol, with a low overhead and designed to the simple, single-hop broadcast communication. CSMA/CA is based on a random backoff algorithm in case the medium is busy, meaning unbounded channel access delays can occur. Adding to that, fairness and scalability are not guaranteed, since some nodes may have to drop several consecutive transmission attempts, particularly in high density scenarios when several nodes simultaneously try to access the medium. Due to that, some nodes might never get access to the medium before the deadline, whereas other nodes might have few difficulties in accessing the medium. Since there is a limited discrete number of backoff slots, it might happen for high density networks, that nodes choose the same number of backoff slots when they sense the medium busy, which might cause a simultaneous transmission within radio range of each other, causing an impact on scalability. This happens particularly in high density and high speed scenarios, such as suburban motorways, for example.

There is the need of a reliable communication infrastructure that can detect safety events and disseminate safety warnings in a secure manner, while being compliant with the maximum latencies involved in safety applications. The communication paradigm can be based on the road side infrastructure (I2V) or to be based on vehicle to vehicle communication (V2V), also known as ad-hoc networks. Hybrid approaches are also possible, particularly in scenarios where vehicle density can vary from traffic jam at certain hours of the day to very few vehicles in the evening. A pure ad-hoc network is quite difficult to manage, and it is likely that vehicle owners will place more trust in a vehicle communication network that is managed by the motorway infrastructure, which can have a global vision of the motorway or at least part of it.

This paper proposes a MAC protocol, based on infrastructure to vehicle communications, that can support safety applications in vehicular environments with a high number of vehicles. In Sect. 2 several proposals of MAC protocols that rely on infrastructure to vehicle communications are presented, and even though they improve the regular IEEE802.11-2012 MAC protocol, they all suffer from one or several shortcomings and therefore there is the need for other proposals. In Sect. 3, the V-FTT protocol is presented and detailed with realistic parameters, proving that it can provide a bounded delay even in worst-case scenarios. Conclusions and future work directions are presented in Sect. 4.

2   Medium Access Control (MAC) in Vehicle to Infrastructure Communications

Most of the MAC protocols are designed to achieve maximum throughput, but vehicular safety applications require small latencies, high data rates and most important require a deterministic delay, meaning that it must be possible to compute the worst-case transmission time. In case a safety application deadline is missed, it might increase the risk of accident, which is much more troublesome than simply degrading the communication system performance.

Several proposals that deal with the MAC issues discussed in the previous section are presented next, with focus on the proposals that use infrastructure-to-vehicle communications.

2.1 Proposals to Address MAC Issues in Dense Vehicular Scenarios

This sub-section presents recent proposals that aim to respond to the MAC layer issues that IEEE802.11-2012 and ETSI G5 suffer in certain scenarios. Böhm and Jonsson [3] introduced a real-time layer on top of the wireless communication standard, at the time IEEE802.11p. They create a super frame with a regular Contention Based Period (CBP) and a Collision Free Phase (CFP). The RSUs assume the responsibility of scheduling the data traffic. They assign each vehicle an individual priority, based on the overall traffic density, its geographical position, since some spots are more dangerous than others, and its proximity to potential hazards. RSUs use a polling mechanism to request vehicles to send their data, which includes position, speed, etc. The super frame begins with a beacon mark which is sent by the RSUs. This beacon informs the duration of the CFP, which can be variable, in order to ensure that all deadlines are met. After this polling phase, every vehicle must switch to the regular contention period (CBP), which can be the IEEE 802.11 MAC mechanism described earlier. Since RSUs need to know exactly which vehicles are inside their communication range, vehicles must send out connection setup requests whenever they hear the RSU beacon. This registration process is done in the CBP, which means that in some cases vehicles might fail to register. Their proposal does not mention how RSU coordinate their beacon transmissions.

The idea of dividing the transmission time into a contention free period and a contention based period was already proposed by Tony Mak et al. [4], by suggesting a change in the 802.11 Point Coordination Function (PCF) mode, in order to adapt it to vehicle communications. They proposed a control channel, where time is divided in periodic intervals (also named repetition period). During the contention free period the RSU poll each vehicle, just as it is done in the IEEE802.11 PCF. The polling list must be updated regularly, which implies that vehicles must register and deregister themselves with the RSUs, using a particular time interval for that purpose. This time interval is announced in a beacon sent by the RSUs in the contention period, meaning that there is a risk that the beacon might not be sent if a large number of nodes tries to access the medium. Knowing that, the authors propose that the beacon is repeated in order to increase the probability of being sent.

Another interesting protocol that was not thought for vehicle communications is Self-organizing TDMA (STDMA). This protocol has similarities with the previous ones: it divides the available time into fixed time slots and organizes them in frames. STDMA is in fact a system in commercial use for ships for transmission collision avoidance. The ships listen to the frame and can determine which slots are free or occupied. Each ship will transmit their position message (the higher the speed, the higher will be the update rate). In case no slots are free, different ships can occupy the same slot according to their position, i.e., a ship will transmit in the same slot as the furthest ship away from itself, in order to reduce interference. STDMA always provides channel access with a bounded delay, and it is scalable. Proposals were made to adapt STDMA for V2V communications [5], concluding that such adaptation requires tight synchronization. The packed drop probability was lower using STDMA than the regular CSMA/CA [5].

In all the protocols presented above, there are critical aspects that were not addressed. The protocols need to be scalable, and there is the need to consider the presence of vehicles that might not be protocol compliant, particularly the ones who use the regular IEEE802.11 MAC protocol. Another issue may occur if RSUs transmission range overlap, which might be needed to ease handover process, and if not considered, it may cause a vehicle to receive different time slots from different RSUs. In order to address these issues, another MAC protocol is proposed in the next section: the vehicular flexible time-triggered (V-FTT), adopting a time division multiple access (TDMA) scheme, where RSUs coordinate themselves to schedule the vehicle transmissions.

3   Vehicular Flexible Time Triggered Protocol

To deal with infrastructure to vehicle (I2V) communications in scenarios where a large number of vehicles, that travel at high speed, want to access the medium of communication within a specific deadline, the Vehicular Flexible Time Triggered protocol (V-FTT) was proposed in [6], inheriting its properties from the Flexible Time Triggered Protocol, which was originally designed for wired real-time communications [7]. Since it might be expensive to supply an entire motorway with a communication network consisting of road side units, a concept of safety zone was devised, meaning that at least the most accident-prone spots of the motorway are covered by RSUs that implement the V-FTT protocol. The Safety Zone concept is depicted in Fig. 1.

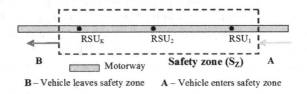

Fig. 1. Safety zone proposal

Inside the Safety Zone, the roadside infrastructure has control of the medium and uses the V-FTT protocol as will be explained next.

3.1   V-FTT Protocol Description

The V-FTT protocol timeline, similarly to the FTT protocol, is divided into elementary cycles (EC). For the specific case of the V-FTT protocol, the EC consists of three time windows:

- The Infrastructure Window (IW) – In this window the RSUs broadcast a Trigger Message (TM) that contains all identifiers of the vehicle on-board units (OBUs) that will be allowed to transmit safety messages in the next window of OBU transmission, the Synchronous OBU Window. Any safety information must be, however, validated from the infrastructure side, and in case it is confirmed, RSUs have specific slots after the TM to send warning messages (WM) to all vehicles that might be affected (protocol enabled and others). The WM have variable duration, depending on the number of occurred events. Each RSU has a fixed size transmission slot, where it transmits its TM and any WMs needed. It is important to notice that there is no medium contention during the IW. For that to be possible, RSUs coordinate their transmissions in the Infrastructure Window so that their transmissions do not overlap. More details can be found in [6].

- The Synchronous OBU Window (SOW) – it is a variable duration window, in which OBUs have the opportunity to transmit their vehicle information (position, speed, etc.) and any safety event (e.g. malfunction or crash warning) based on the vehicle on-board sensors. For that purpose, they have a fixed size slot (SM), which was assigned in the previous IW. This means they can transmit without medium contention. To ensure fairness in medium access, each vehicle is only allowed to have one transmission slot per SOW.

- The Free Period Window (FP) – This is in fact a contention based window, where all vehicles that do not follow the V-FTT protocol (non-enabled OBUs) are able to contend for the medium, usually for transmission of short messages unrelated to safety. The enabled OBUs are also allowed to contend for the medium but do not have any transmission guarantees. It is important to guarantee that a FP exist in most of the EC in order to allow non-enabled vehicles to transmit their information.

Figure 2 presents the Elementary Cycle and its three transmission windows.

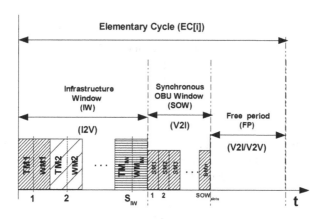

Fig. 2. V-FTT protocol Elementary Cycle

The V-FTT protocol fits comfortably on top of the current wireless standards devised for vehicular communications, IEEE802.11-2012 or ETSI-G5, where in the first case the duration of the Elementary Cycle matches the duration of the Control Channel (CCH) Interval (100 ms), and in the latter, it can vary according to the need.

The next sub-section summarizes the proposal validation, that was made by quantifying several protocol parameters in worst-case scenario.

3.2 Protocol Parameters

In order to assess the V-FTT applicability to real scenarios using current wireless communication standards, several protocol parameters need to be quantified [8]. The maximum range of communication of the IEEE802.11-2012 standard is 1000 m [9], but 750 m have been proven to be a more cautious figure [10], meaning that the RSU coverage range C_r is assumed to have a value of 750 m. In order to ease the handover process in a high speed scenario such as vehicle environment, the overlap of RSU coverage O_r is assumed to be at least 25% of the coverage range [11]. This allows to determine the overlapping range O_r as well as the maximum spacing between RSUs. The vehicle average length and average spacing are based on [12, 13]. All parameters are summarized in Table 1.

Table 1. Road side units and vehicle parameters.

Parameter acronym	Parameter	Value (m)
C_r	Coverage range …	750
O_r	Overlapping range	187,5
S_r	Maximum spacing between consecutive RSUs	1312,5
V_{length}	Average vehicle length	4,58
$V_{spacing}$	Traffic jam average vehicle spacing	10
$V_{spacing}$	Normal traffic average vehicle spacing	30

Table 2. Maximum number of vehicles covered per lane by an RSU with 750 m of coverage.

Lanes per travel path	Normal traffic	Traffic jam
1	44	103
4	174	412
5	217	507

According to the values shown in Table 1, considering that vehicle average length (V_{length}) is 4,58 m, it is straightforward to compute the maximum number of vehicles covered by each RSU, which is dependent on the number of lanes of the motorway, as shown in Eq. (1):

$$N_{VRSU} = \frac{2 \times C_r}{\left(V_{length} + v_{spacing}\right)} \times n_{lanes}. \tag{1}$$

Considering that the Safety Messages (SM) must include several fields such as message identifiers, time stamps, vehicle data (position, speed, acceleration, etc.) and safety events warning, it was shown in [6] that its minimum size is 390bit, for non-encrypted data. Assuming that this SM is transmitted using a similar physical layer than IEEE-802.11-2012 or ETSI-G5 then for an OFDM 10 MHz channel its transmission duration is dependent on the bit rate used. The maximum available number of transmission slots for V2I transmission without contention (Synchronous OBU Window) was computed in [8], for the case the Elementary Cycle matches the value of the CCH Interval duration in the IEEE802.11-2012 standard, which is 100 ms. Results are shown in Table 3.

Table 3. Maximum number of transmission slots without contention per Elementary Cycle

Bit rate (Mbps)	SOW transmission slots (no Free Period allowed)	SOW transmission slots (when FP = 10% of EC)
3	281	251
6	473	424
12	688	618

Analysing the value in Table 3 and those of Table 2 it is straightforward to conclude that it might be worth to suppress the Free Period in some exceptional cases, only during a small amount of time, to allow more vehicles to communicate in the SOW. Another important conclusion is that there is the need of a scheduling mechanism that can fairly allocate vehicle transmissions in the slots of the OBU window.

4 V-FTT Worst Case Analysis

In order to validate the proposed V-FTT protocol two different scenarios were considered: normal traffic conditions, where the average distance between vehicles is 30 m, and traffic jam, where the average distance was considered to be 10 m [13]. A fairness condition was imposed to the scheduling mechanism of the OBU communications, meaning that after the RSU attribute a time slot for a specific OBU, its next opportunity of transmitting without contention will only occur after all other OBUs in the Safety Zone have had their time slot attribution. An important evaluation parameter is the worst case delay in what concerns the amount of time that occurs between an event detection until the last vehicle in the safety zone is warned. This is named t_{worst}.

For the V-FTT protocol the involved times in the process are:

- t_{V2I} – period of time that occurs after a vehicle detects an event until it can effectively transmit that information to an RSU.
- t_{I2V} – period of time that occurs after a RSU schedules a warning message (WM) until it is effectively transmitted.

In fact, there are other times involved, since the infrastructure must validate the event detected by an OBU, and must schedule the transmission of a WM. For a worst case analysis these times are negligible when compared to the other times involved, and therefore they are not considered.

The worst-case for t_{V2I} occurs when a vehicle detects the event just after its last transmission in its respective time slot in the SOW, meaning that the vehicle must wait until its next transmission opportunity. For reasoning purposes, this vehicle is named emitter vehicle. With the scheduler fairness restrictions, the worst case scenario will occur when the emitter vehicle is only allowed to transmit after all the remaining vehicles have had their transmission opportunity (at least the vehicles that are present in the same coverage area of the Safety Zone). Since the number of vehicles can have large variations, the worst-case is considered to occur when the Safety Zone is fully populated with vehicles. This number depends on the number of lanes of the motorway, since more lanes imply that more vehicles can be present. If the number of vehicles travelling in the Safety Zone exceeds the number of transmissions slots available in the SOW, it might happen that the emitter OBU must wait for more than one EC for its transmission opportunity. This is referred as number of waiting elementary cycles w_{EC} (please refer to Fig. 3). Assuming that scheduling is made in every EC, the emitter vehicle is at least certain that its transmission slot will occur in the SOW after w_{EC}. It might occur in the first slot or the last slot of the SOW after w_{EC}, the latest being the worst-case, which results in the equation shown in (2):

$$t_{V2I} = SOW + (w_{EC} + 1) \times E, \tag{2}$$

where SOW is the duration of the Synchronous OBU Window, and E is the duration of an Elementary Cycle.

The worst case t_{V2I} value is exemplified in Fig. 3.

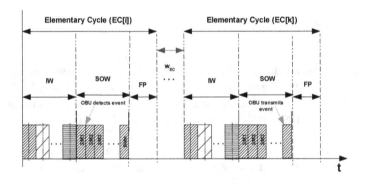

Fig. 3. Worst case vehicle transmission instant (t_{V2I})

As for the case of the downlink time, the RSU always receives vehicle information in the SOW, meaning it has to wait for the next IW (in the next EC) to transmit any warning message. The worst case occurs if the information received from the vehicle is sent in the beginning of the SOW, which means the waiting time will be maximum, assuming that the validation of the safety event does not interfere with the involved times here described. It was shown in [8] that the duration of the SOW can vary, with its maximum value occurring when no free period is allowed. In other words, the maximum value of t_{I2V} is the full length of an Elementary Cycle (E), minus the duration of the TM used by the vehicle to convey the safety event information (please refer to (3)).

$$t_{I2V} = (E - TM). \tag{3}$$

After determining t_{V2I} and t_{I2V}, one might think that t_{worst} is obtained by simply adding the first two parameters, however those worst uplink and downlink times never occur simultaneously [8]. Therefore, assuming that the value of the Free Period has negligible variations from one elementary cycle to the next, and also assuming that the duration of the EC remains constant during the period of time involved, it can be proved that t_{worst} is determined by (4):

$$t_{worst} = (w_{EC} + 2) \times E. \qquad (4)$$

Looking at Eq. (4), it seems that t_{worst} is linearly dependent on the duration of the Elementary Cycle, and that a reduction on the EC length can provide better results. However, such measure would reduce the length of SOW and consequently the number of vehicle transmission slots per EC, which might increase the value of w_{EC} for scenarios with a high number of vehicles, thus inflating the value of t_{worst}. The values for t_{worst} are summarized in Table 4.

Table 4. Worst case warning time (no FP and fair scheduling, Safety Zone with 4 lanes per travel path)

Bit rate (Mbps)	Normal traffic	Traffic jam
3	300 ms	400 ms
6	200 ms	300 ms
12	200 ms	300 ms

Analysing Table 4, it can be seen that the worst-case results for the lowest bit rates do not allow to support some of the safety applications that need to have low latencies. On the other hand, for the case of the traffic jam scenario, vehicles are not expected to travel at very high speeds, which can increase the maximum latency allowed for safety applications. For the highest bit rates, this problem does not occur. Most importantly, the V-FTT protocol provides a maximum bounded delay even in worst-case scenarios. A disadvantage of using a wireless communication standard that has a fixed EC (CCH interval), such as the IEEE802.11-2012, is that it is not possible to reduce the duration of the EC, which could provide better worst-case values for the cases where the number of vehicles can be scheduled in one SOW.

5 Conclusions and Future Work

This paper discussed an existing shortcoming in current wireless standards proposed for vehicle communications: in certain scenarios, such as high speed motorways with a high number of OBUs, the MAC layers of these standards do not offer a guaranteed bounded delay, which can pose a problem for the deployment of safety vehicle applications with low latency needs.

Several proposals to address these shortcomings were discussed, particularly those based on an infrastructure to vehicle communication, but the lack of RSU coordination and the fact that most of the proposals do not allow OBUs that are not compliant to their protocol lead to the proposal of the Vehicular Flexible Time Triggered protocol (V-FTT), an infrastructure based communication protocol.

In order to demonstrate that the V-FTT protocol has a maximum bounded delay, a worst-case scenario was defined, in terms of the maximum amount of time that occurs between an event detection until all vehicles in the safety zone are warned. It was shown that indeed a bounded delay is obtained, although its value when using the lowest available bit rate is not enough for some safety applications, which in turn leads to the conclusion that a scheduling mechanism is needed to reduce the times involved, while maintaining fairness and scalability of the protocol.

Future work involves testing a real-time scheduler with other parameters, in order to analyse the V-FTT protocol in different scenarios using different algorithms such as Earliest Deadline First or rate monotonic scheduling. Simulation will also be used to compare the behaviour of the V-FTT protocol with the regular MAC layer of IEEE 802.11-2012 and other solutions.

References

1. IEEE. 802.11-2012 - IEEE Standard for Information technology–Telecommunications and information exchange between systems Local and metropolitan area networks–Specific requirements Part 11: Wireless LAN Medium Access Control (MAC) and Physical Layer (PHY) Specifications (2012)
2. ETSI ITS-G5 standard - Final draft ETSI ES 202 663 V1.1.0, Intelligent Transport Systems (ITS); European profile standard for the physical and medium access control layer of Intelligent Transport Systems operating in the 5 GHz frequency band (2011)
3. Böhm, A., Jonsson, M.: Real time communications support for cooperative, infrastructure-based traffic safety applications. Int. J. Veh. Technol. **2011** (2011)
4. Mak, T., Laberteaux, K., Sengupta, R.: A multi-channel VANET providing concurrent safety and commercial services. In: VANET 2005, Germany, September 2005
5. Bilstrup, K., Uhlemann, E., Ström, E., Bilstrup, U., On the ability of the 802.11p MAC method and STDMA to support real-time vehicle-to-vehicle communication. EURASIP J. Wireless Commun. Networking **2009**, Article ID 902414 (2009)
6. Meireles, T., Fonseca, J., Ferreira, J.: The case for wireless vehicular communications supported by roadside infrastructure. In: Perallos, A., Hernandez-Jayo, U., Onieva, E. (eds) Intelligent Transportation Systems Technologies and Applications. John Wiley and Sons (2015)
7. Pedreiras, P., Almeida, L.: The Flexible Time-Triggered (FTT) paradigm: an approach to QoS management in distributed real-time systems. In: Proceedings of International Parallel and Distributed Processing Symposium (2003)
8. Meireles, T., Fonseca, J., Ferreira, J.: Deterministic vehicular communications supported by the roadside infrastructure. In: Alam, M., Ferreira, J., Fonseca, J. (eds) Intelligent Transportation Systems. SSDC, vol. 52, pp. 49–80. Springer, Cham (2016)
9. Gallagher, B., Akatsuka, H.: Wireless communications for vehicle safety: radio link performance & wireless connectivity methods. IEEE Veh. Technol. Mag. **1**, 4–24 (2006)

10. Stibor, L., Zang, Y., Reumerman. H.: Evaluation of communication distance of broadcast messages in a vehicular Ad-Hoc network using IEEE 802.11p. In: Wireless Communication and Networking Conference (WCNC 2007), Hong Kong (2007)
11. Ancusa, V., Bogdan, R.: A method for determining Ad-Hoc redundant coverage area in a wireless sensor network. In: 2011 2nd International Conference on Networking and Information Technology, IPCSIT, vol. 17 (2011)
12. Crow, A.: Recommendations for traffic provisions in built-up areas (1998)
13. Xu, Q., Sengupta, R., Mak, T., Ko, J.: Vehicle to vehicle safety messaging in DSRC. In: 2004 Proceedings of the 1st ACM International Workshop on Vehicular Ad Hoc Networks, VANET 2004 (2004)

A Proposal for an Improved Distributed MAC Protocol for Vehicular Networks

Aqsa Aslam[1($\boxtimes$)], Luis Almeida[1], and Joaquim Ferreira[2]

[1] Instituto de Telecomunicações/FEUP, Universidade Do Porto, Porto, Portugal
engr.aqsa.tl@gmail.com, lda@fe.up.pt
[2] Instituto de Telecomunicações/ESTGA, Universidade de Aveiro, Aveiro, Portugal
jjcf@ua.pt

Abstract. Vehicular Ad-hoc Networks (VANETs) have a significant potential to enable new applications in the vehicular domain, some of which addressing traffic safety. In these networks, the Medium Access Control (MAC) plays an important role in providing an efficient communication channel. Currently, there are two protocols for ITS, proposed in the USA (IEEE WAVE) and in Europe (ETSI ITS-G5). Both cases use the PHY and MAC of IEEE 802.11p, the latter being fully distributed and based on CSMA/CA, thus still prone to collisions. This has led to recent proposals for TDMA-based overlay protocols to prevent collisions but leading to complex synchronization and scalability limitations. In this paper we propose enhancing IEEE 802.11p with an overlay protocol based on Reconfigurable and Adaptive TDMA. We specifically target multiple concurrent applications such as multiple platoons. Our proposal separates between the application level, with its own TDMA round and slots allocated to engaged nodes, e.g., a platoon, and the global level that manages multiple TDMA rounds in a mutually agnostic manner, thus, dynamic and scalable. We believe this is the first applications-oriented MAC protocol proposed for VANETS and we discuss its deployment and potential advantages in two typical uses cases, namely platooning and smart intersections.

Keywords: VANET · IEEE 802.11p · TDMA · Admission control

1 Introduction

The impact of increasing road accidents and traffic congestion has motivated the appearance of Intelligent Transportation Systems (ITS) and associated applications that aim at improving road efficiency and safety [1]. Vehicular Ad-hoc Networks (VANETs) are important components of an ITS in which all vehicles are equipped with wireless devices that support collaborative applications, be it for safety purposes (such as accident alerts) or non-safety (such as Internet access), through direct Vehicle-to-Vehicle (V2V) or Vehicle-to-Infrastructure (V2I) communication.

© ICST Institute for Computer Sciences, Social Informatics and Telecommunications Engineering 2017
J. Ferreira and M. Alam (Eds.): Future 5V 2016, LNICST 185, pp. 24–33, 2017.
DOI: 10.1007/978-3-319-51207-5_3

These collaborative applications generally benefit from a more reliable communication channel with less access collisions, lower latency and higher throughput, properties that call upon an adequate Medium Access Control (MAC) protocol. Currently, there are two similar but non-interoperable ITS communication standards, operating in the 5.9 GHz band, namely the IEEE Standard for Wireless Access in Vehicular Environments (WAVE) and the ETSI Standard for Intelligent Transport Systems (ITS-G5). At the physical and MAC layer, both use the IEEE 802.11p protocol to arbitrate access within each of the channels provided by the standards, both control and service channels. This protocol relies on the well known CSMA/CA distributed access arbitration method, with different enhancements on both cases. Nevertheless, IEEE 802.11p still suffers from chained collisions and poor performance under dense traffic situations [2,3].

Recently, several MAC protocols that are based on the Time Division Multiple Access (TDMA) technique were proposed to allow vehicles to use the same frequency channel without, or with less, transmission collisions. This technique divides time in consecutive non-overlapping slots and allocates each slot to one vehicle for exclusive channel access. In the specific case of VANETs, vehicle mobility must be considered since it causes the network topology to change, particularly regarding number of engaged nodes. Thus, the TDMA mechanism must also provide dynamic slot assignment [4].

As we will show in the following section, the issue of avoiding collisions at the MAC layer in an efficient way is still open. Thus, in this paper, we propose a new MAC overlay protocol for use over IEEE 802.11p. In particular, we aim at improving the channel quality provided to each well defined set of interacting vehicles, i.e., those engaged in a specific collaborative application, using TDMA to reduce access collisions. Moreover, we propose a mechanism that allows multiple TDMA rounds to co-exist in time and space, corresponding to concurrent groups of interacting vehicles, i.e., concurrent collaborative applications. Thus, we aim at providing per application slots coordination (slots in one TDMA round), together with coordination of multiple applications (multiple TDMA rounds).

For this purpose, motivated by the work done in [5] we will make use of the dynamic mechanisms of the Reconfigurable and Adaptive TDMA (RA-TDMA) protocol, which was developed to provide communications for dynamic teams of cooperating mobile autonomous robots. With RA-TDMA we can offer a dynamic fully distributed TDMA framework on top of IEEE 802.11p to combine the benefits of both TDMA and CSMA/CA paradigms, namely collision reductions with efficient bandwidth usage. Moreover, the reconfiguration feature of RA-TDMA can handle the slots coordination per application while the adaptation feature can handle the co-existence of multiple applications by adjusting the phases of the respective TDMA rounds.

To the best of our knowledge, this paper puts forward the first proposal towards an application-oriented TDMA overlay MAC protocol and is organized as follows. The next section discusses related work. Section 3 introduces the basics of RA-TDMA. Our proposal is presented and discussed in Sect. 4 while

Sect. 5 presents two illustrative use cases. A concluding Sect. 6 wraps up the paper and discusses the next steps in our work.

2   Related Work

Several methods are available in the literature for the MAC protocol in VANETs, particularly in the context of road safety. Generally, MAC protocols are classified in two broad categories such as contention-based and schedule-based. In contention-based MAC protocols, each node tries accessing the channel when it has data to transmit using the CSMA mechanism. IEEE 802.11p, used by both ITS standards, is a contention-based MAC protocol. In WAVE, it is further enhanced with the priority based access scheme of Enhanced Distributed Channel Access (EDCA). Conversely, in ITS-G5 it is enhanced with Distributed Congestion Control (DCC) which acts on certain MAC parameters (e.g., transmission frequencies, data rate and power levels). However, both enhancements do not preclude access collisions and the quality of the channel can be significantly degraded in the presence of intense localized traffic, particularly of the same priority class [3,6,7].

Thus, several recent works have proposed using TDMA-based MAC techniques that either fully eliminate or at least significantly reduce access collisions [1,8]. At the VANET level, the benefits of these techniques include a fair (equal) access to the channel for all vehicles, improved reliability of vehicles communications, more efficient channel utilization due to less collisions, deterministic network access time under dense traffic and QoS suitable for real-time applications. Among these techniques, some use centralized traffic scheduling in the Road-Side Units (RSU), such as V-FTT [9] to achieve superior traffic management capabilities and meeting strict real-time guarantees. Conversely, in this work we follow a fully distributed approach that does not necessarily require RSUs, even if providing weaker real-time guarantees, thus leading to lower infrastructure requirements and easier deployment. Several protocols were proposed in this direction that are particularly related to our work and which we discuss next.

VeMAC [10] is a contention-free protocol that supports efficient broadcast in the control channel. This protocol reduces collisions by assigning disjoint sets of time slots to vehicles moving in opposite directions (Left and Right) and to the road side unit (RSU).

DMMAC [11], standing for Dedicated Multi-channel MAC protocol, has an adaptive broadcasting mechanism designed to provide collision-free and delay-bounded transmissions for safety applications under different traffic conditions. It divides the control channel into an adaptive broadcast frame that further consists of time slots and each time slot is reserved by one vehicle for collision free transmission of safety messages.

DTMAC [12], standing for fully Distributed TDMA-based MAC protocol, extends VeMAC with a traffic scheduling approach that considers the road dissected into small fixed areas in which the time slots can be reused. Thus, it contributes to alleviate the scalability limitations of VeMAC by allowing parallel transmission in different areas.

VeSOMAC [13] aims at achieving fast TDMA slot reconfiguration without relying on roadside infrastructure for coping with vehicular topology changes. The allocation scheme is based on a bitmap in-band signaling scheme that carries information about allocated slots and allows fast slot reconfiguration following topology changes such as when platoons merge.

STDMA [14] presents a decentralized TDMA scheme aiming at real-time communication. It uses periodic frames further divided in time slots. When a vehicle joins the VANET, it first listens to the channel to get information from other vehicles positions and then performs four different phases, namely initialization, network entry, first frame, and continuous operation. This approach reserves slots for the following transmissions to provide exclusive access to the channel.

All previous approaches consider the communication channel as a global entity that is partitioned in time slots in different ways. This raises a scalability issue, limiting the number of vehicles that can engage the VANET. However, some of the approaches already include mechanism to overcome this limitation, such as DTMAC and STDMA, but both with limited efficiency given their specific slot reuse techniques. Moreover, all those approaches use complex synchronization mechanisms to virtually avoid slots overlapping and transmissions collisions. In particular, they do not generally use the underlying CSMA/CA native MAC in IEEE 802.11p to handle possible asynchronous transmissions.

Conversely, our proposal follows a significantly different approach based on RA-TDMA [5]. It synchronizes small sets of vehicles, only, i.e., those engaged in each collaborative application, making them transmit in a round with an adequate period and transmissions separated in time as much as possible. Simultaneously, our proposal tolerates asynchronous traffic using the CSMA/CA native MAC. Moreover, our approach uses a fully distributed adaptive mechanism that allows multiple synchronized rounds, associated to different collaborative applications, to coexist intermingled. Such mechanism keeps the rounds out of phase using a feedback approach based on the detection of mutual interference that they can occasionally cause to each other. As a result, there is no concept of slot reuse and the whole channel can be reused up to its capacity. When one application is within the range of another one, the phase of its round will be adjusted as needed to avoid the interference. As an application moves away from another one, its interference ceases and parallel transmission can occur without any adjustment, leading to a full channel reuse.

To the best of our knowledge, this is a novel approach to the joint management of a shared channel by multiple concurrent applications that set up a synchronous set each. Given the full channel reuse and absence of global management structures, we claim that our approach provides full scalability.

3   RA-TDMA Basics

We will build upon our previous experience with dynamic teams of autonomous mobile robots making use of the RA-TDMA protocol that was explicitly developed for that context [5]. Thus, in this section we review its main features.

RA-TDMA is an overlay protocol that works on top of a native distributed arbitration mechanism, typically CSMA/CA, such as in IEEE 802.11. The nodes engaged in the protocol form a team and transmit periodically in a round, with a predetermined period that meets the application requirements (t_{up}). The round is divided into a dynamic number of slots according to the current number of nodes in the team. Figure 1a depicts the structure of two consecutive RA-TDMA rounds where t_{r_1} represents the beginning of round 1 and $t_{s_{1,0}}$ represents the beginning of slot 0 of round 1. The nodes transmit as soon as possible in their slots, thus separating the transmissions of team nodes in time as much as possible. The synchronization of all team nodes in a common round is achieved trying to enforce a slot separation between consecutive transmissions, thus without using clock synchronization (Fig. 1b).

If other nodes are not engaged in the team, or when new nodes try to join the team, the underlying CSMA/CA arbitration is used to control access to the channel. This can cause delays in the regular team transmissions. However, by trying to keep a slot separation between consecutive team transmissions the protocol incorporates such delays resulting in a phase adjustment of the TDMA round. This mechanism effectively avoids periodic interference as that caused by other nodes transmitting periodically with similar periods. Figure 1b shows the synchronization mechanism where the initial slots are marked with dashed lines. A delay in node 0 is noticed by node 2 that delays its next slot setting a new time-frame, marked with full lines. Nodes 1 and 3 are still unaware of this delay and keep their initial slots. Once node 2 transmits in the adjusted slot, node 1 is made aware of this adjustment and will synchronize. Finally, node 3 will also synchronize after receiving a packet from node 1.

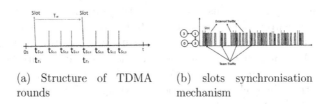

(a) Structure of TDMA rounds

(b) slots synchronisation mechanism

Fig. 1. RA-TDMA round with synchronization mechanism

4   Towards an Application-Oriented VANET MAC

In this paper, we propose using RA-TDMA within the vehicular context to manage state sharing in the scope of collaborative applications.

In this case, each group of vehicles engaged in one collaborative application, e.g., a platoon or smart intersection, form a team. RA-TDMA is then used to manage the periodic transmissions of each vehicle in the team for state sharing, allocating them to different slots in a TDMA round with an application dependent fixed period.

As typical in RA-TDMA, the slots will be larger than the nodes communication requirements, leading to substantial available time between transmissions of consecutive nodes in the round. These periods of availability are used to tolerate external traffic, i.e., traffic not engaged in the application, using the underlying CSMA/CA native MAC.

We propose using these *availability periods* to support the coexistence of multiple applications, each with its own TDMA round and considering all other as external traffic. The phase adaptation embedded in RA-TDMA allows setting the multiple rounds out of phase without being explicitly aware of each other. Each application (round) simply feels the delays in its own traffic caused by the interference of the other rounds.

We conjecture that such coordination of transmissions enhances channel Quality-of-Service (QoS) in an inherently scalable manner.

4.1   Application-Based Admission Control

One aspect that needs to be addressed when applying RA-TDMA in the scope of VANETs is Admission Control. In the original RA-TDMA, any robots within communication reach of the team would automatically be incorporated. However, with vehicular coordination applications, it is not enough to consider the vehicles that are within communication range but also other application dependent aspects, such as vehicles position and application capacity. For example, a platooning application may be limited to control a maximum of 5 vehicles, or a smart intersection may be set to coordinate a maximum of 20 vehicles. In such cases, there may be more vehicles within the communication range, thus, those that must be integrated in the team are those that are in more adequate physical positions, such as consecutive aligned positions in platoons or those that are closer to the intersection core in smart intersections.

Therefore, our proposal also includes adding an admission control module for RA-TDMA that verifies (i) whether the application vehicles capacity is exhausted and (ii) whether the position of the joining vehicle is compatible with the application. Moreover, it is likely that vehicles will engage one application at a time, e.g., only one platoon, or one intersection.

Figure 2 shows a simplified diagram with the Admission Control (AC) deciding whether a given vehicle can be incorporated in the respective application RA-TDMA round or whether it will be rejected and its transmissions considered as external traffic and applied directly to the underlying protocol.

Note that this admission control approach does not prevent nodes from transmitting. In fact, when a join request is rejected, the requesting application should back off and retry later on, if convenient. The vehicle can continue issuing other joining requests, potentially joining another application. It is also common that the frequency with which a vehicle shares its state is lower before engaging in an application (asynchronous scanning phase) and faster once accepted (synchronous collaboration phase).

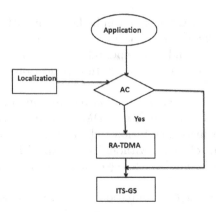

Fig. 2. Adding Admission Control (AC) on top of RA-TDMA

5   Illustrative Use Cases

In this section we present two use cases that show how our proposal works in practice. One is a platooning scenario and another one is a smart intersection.

5.1   Platooning Scenario

In long distance roads, particularly highways, groups of vehicles can coordinate to travel closely together in a straight line thus improving safety, fuel economy, driver comfort while reducing traffic congestion. These groups of vehicles traveling together at approximately constant speed and relatively short inter-vehicle distance are called platoons. These formations have one leader and a number of followers that is typically bounded. For control purposes, their speeds and positions are exchanged at a relatively high frequency. A vehicle can only be a member of one platoon at a time.

Consider that we have a scenario like the one shown in Fig. 3a. Platoon A consists of three cars in the middle lane that are already engaged in such collaboration. Meanwhile another vehicle approaches from the tail of the platoon, with a platoon application enabled. At a certain moment it starts hearing the platoon messages. At that moment it checks if it is compatible with that platoon and issues a join request. When this request is received by at least one platoon member, such member shares the request with the whole platoon and invokes the admission control that will take a consistent decision. If the on-going platooning application can accept a fourth vehicle and it is in a compatible position, i.e., inline with the platoon and at the tail, at less than a maximum distance, then the application will reconfigure its round to create a new slot and the new vehicle will integrate the platoon. If the maximum number of vehicles was four, a fifth vehicle requesting to join would be rejected from this platoon but it still continuous transmitting as external traffic.

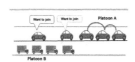

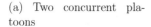

(a) Two concurrent platoons

(b) Two concurrent RA-TDMA rounds

Fig. 3. Platooning use case with multiple RA-TDMA rounds

To highlight the position-based feature of our proposed admission control, consider now the situation in which two vehicles approach platoon A at approximately the same time (Fig. 3a) and request joining. If both requests are considered together, the admission control should favor the vehicle closer to the platoon and reject the farther one. If for some reason, e.g., a better antenna, the last vehicle issues a join request before, the admission control should detect the large distance to the current platoon tail and reject it, later accepting the joining request from the closer vehicle.

Finally, when two different platoons meet each other, such as platoons A and B in Fig. 3a, they check whether they are compatible, i.e., same type of vehicles on the same lane. If so, if the total number of vehicles is within the capacity of at least one of the platooning applications and the platoons are sufficiently closer to each other, both platoons are merged. However, if any of these conditions fails, the platoons continue separately and their RA-TDMA rounds are interpreted as external traffic to each other. This is the major feature of our proposal that grants it full scalability.

5.2 Intersection Scenario

Another useful traffic coordination scenario is that of smart intersections where the vehicles automatically coordinate to arbitrate access to the shared area, possibly using an intersection controller as shown in Fig. 4a. The intersection controller then issues announcing messages that will be detected by the vehicles within range. These will check whether they are approaching or leaving the intersection and, if approaching, issue joining requests to the intersection application. Again, the admission control feature will check the capacity of the application and the position of the joining nodes, accepting those that are sufficiently close to the center of the intersection, only. Moreover, if the capacity limit is reached and a joining request comes from a vehicle closer to the intersection than another one already engaged in the application in the same lane, the latter is disengaged and the request accepted. Similarly, when a vehicle crosses the intersection, it is disengaged, too, giving room for approaching vehicles.

Finally, when a platoon reaches a smart intersection, the latter gains priority and the platoon application disengages all its vehicles that will then compete for

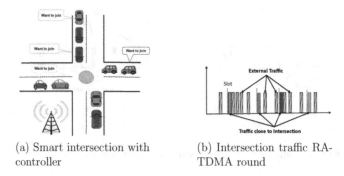

(a) Smart intersection with controller

(b) Intersection traffic RA-TDMA round

Fig. 4. Smart intersection use case with an RA-TDMA round

the access to the intersection application. New platoons can then be automatically formed at the exit of an intersection.

6 Conclusion

In VANETs, the design of efficient MAC protocols is an important issue due to the impact this layer has on the performance of collaborative traffic applications. In this paper we proposed adding an overlay transmissions coordination protocol to the MAC layer of current ITS standards, essentially based on IEEE802.11p. This overlay protocol is based on RA-TDMA, previously developed for dynamic teams of robots. We use RA-TDMA, enhanced with a position-based admission control, to coordinate the traffic issued by vehicles engaged in a particular collaborative application. We use the reconfigurable part of RA-TDMA to automatically create a round with as many slots as active vehicles, keeping their transmissions as separated in time as possible, thus reducing collisions. However, RA-TDMA also tolerates external traffic, handled with the CSMA/CA arbitration of IEEE802.11p.

Particularly, multiple concurrent collaborative applications can now coexist, each with its own RA-TDMA round, seeing the others as external traffic. The adaptive feature of RA-TDMA allows dephasing the multiple rounds to minimize interference among each other. This feature simplifies the protocol management, which is fully distributed, reduces collisions without any global information and uses minimal configuration thus granting full scalability. To the best of our knowledge, this is the first proposal for such an application-level overlay MAC protocol.

In the next steps we will formalize, model and implement the proposed approach, to quantify its benefits.

References

1. Hadded, M., Muhlethaler, P., Laouiti, A., Zagrouba, R., Saidane, L.A.: TDMA-based MAC protocols for vehicular ad hoc networks: a survey, qualitative analysis, and open research issues. IEEE Commun. Surv. Tutorials **17**(4), 2461–2492 (2015)
2. E. ETSI, 302 663 (v1. 2.1) (11-2012): Intelligent Transport Systems (ITS), Access layer specification for Intelligent Transport Systems operating in the 5 GHz range (2012)
3. Eckhoff, D., Sofra, N., German, R.: A performance study of cooperative awareness in ETSI ITS G5 and IEEE WAVE. In: 2013 10th Annual Conference on Wireless On-demand Network Systems and Services (WONS), pp. 196–200, March 2013
4. Chen, Q., Jiang, D., Delgrossi, L.: IEEE 1609.4 DSRC multi-channel operations and its implications on vehicle safety communications. In: 2009 IEEE Vehicular Networking Conference (VNC), pp. 1–8. IEEE (2009)
5. Santos, F., Almeida, L., Lopes, L.S.: Self-configuration of an adaptive TDMA wireless communication protocol for teams of mobile robots. In: IEEE International Conference on Emerging Technologies and Factory Automation, ETFA 2008, pp. 1197–1204. IEEE (2008)
6. Uzcategui, R., Acosta-Marum, G.: Wave: a tutorial. IEEE Commun. Mag. **47**(5), 126–133 (2009)
7. Subramanian, S., Werner, M., Liu, S., Jose, J., Lupoaie, R., Wu, X.: Congestion control for vehicular safety: synchronous and asynchronous MAC algorithms. In: Proceedings of the Ninth ACM International Workshop on Vehicular Internetworking, Systems, and Applications, ser. VANET 2012, pp. 63–72. ACM, New York (2012)
8. Autolitano, A., Campolo, C., Molinaro, A., Scopigno, R., Vesco, A.: An insight into decentralized congestion control techniques for VANETs from ETSI TS 102 687 V1.1.1. In: 2013 IFIP Wireless Days (WD), pp. 1–6, November 2013
9. Meireles, T., Fonseca, J., Ferreira, J.: Deterministic vehicular communications supported by the roadside infrastructure: a case study. In: Alam, M., Ferreira, J., Fonseca, J. (eds.) Intelligent Transportation Systems. SSDC, vol. 52, pp. 49–80. Springer, Cham (2016)
10. Omar, H., Zhuang, W., Li, L.: VEMAC: a TDMA-based MAC protocol for reliable broadcast in VANETs. IEEE Trans Mob. Comput. **12**(9), 1724–1736 (2013)
11. Lu, N., Ji, Y., Liu, F., Wang, X.: A dedicated multi-channel MAC protocol design for VANET with adaptive broadcasting. In: IEEE Wireless Communication and Networking Conference, pp. 1–6. IEEE (2010)
12. Hadded, M., Laouiti, A., Zagrouba, R., Muhlethaler, P., Saidane, L.A.: A fully distributed TDMA based MAC protocol for vehicular ad hoc networks. In: International Conference on Performance Evalutaion and Modeling in Wired and Wireless networks, PEMWN 2015 (2015)
13. Yu, F., Biswas, S.: A self-organizing MAC protocol for DSRC based vehicular ad hoc networks. In: 27th International Conference on Distributed Computing Systems Workshops, ICDCSW 2007, pp. 88–88. IEEE (2007)
14. Alonso, A., Sjöberg, K., Uhlemann, E., Ström, E., Mecklenbräuker, C.: Challenging vehicular scenarios for self-organizing time division multiple access. European Cooperation in the Field of Scientific and Technical Research (2011)

Adaptive Contention Window Design to Minimize Synchronous Collisions in 802.11p Networks

Syed Adeel Ali Shah$^{(\boxtimes)}$, Ejaz Ahmed, Iftikhar Ahmad, and Rafidah MD Noor

Faculty of Computer Science and Information Technology, University of Malaya,
50603 Kuala Lumpur, Malaysia
{adeelbanuri,ify_ia}@siswa.um.edu.my, ejazahmed@ieee.org, fidah@um.edu.my

Abstract. The vehicular ad hoc network (VANET) capable of wireless communication will enhance traffic safety and efficiency. The IEEE 802.11p standards for wireless communication in the US and Europe use a single shared channel for the periodic broadcast of safety messages. Coupled with the short contention window and inflexibility in window size adaptation, the synchronous collisions of periodic messages are inevitable in a large scale intelligent transportation system (ITS). To this end, we propose an adaptive contention window design to reduce synchronous collisions of periodic messages. The proposed design replaces the aggressive window selection behaviour in the post transmit phase of IEEE 802.11p with a weighted window selection approach after a successful transmission. The design relies on the local channel state information to vary contention window size. Moreover, in high density networks, the design gives prioritized channel access to vehicles experiencing dropped beacons. The proposed design can be readily incorporated into the IEEE 802.11p standard. The discrete-event simulations show that synchronous collisions can be reduced significantly to achieve higher message reception rates as compared to the IEEE 802.11p standard.

Keywords: VANET · Synchronous collisions · ITS · 802.11p · Congestion control · Media access control · Contention window · Adaptive contention window

1 Introduction

The research in Vehicular Ad hoc Network (VANET) has received much interest due to its potential to provide drivers not only with safety specific data but with information useful for traffic efficiency and passenger comfort [1–3]. The key concept of transmitting such information is the use of wireless communication technology based on IEEE 802.11p standard [4,5]. The transmission of safety information messages (i.e. beacons) is frequent and valid for a limited time period. It implies that the Medium Access Control (MAC) layer specification in IEEE 802.11p has to fulfill specific requirements for efficient operation of Intelligent Transportation System (ITS).

© ICST Institute for Computer Sciences, Social Informatics and Telecommunications Engineering 2017
J. Ferreira and M. Alam (Eds.): Future 5V 2016, LNICST 185, pp. 34–45, 2017.
DOI: 10.1007/978-3-319-51207-5_4

Due to high frequency of beacons, one crucial requirement is to efficiently utilize the limited available wireless spectrum for reliable beacon delivery. In high density vehicular networks, the amount of periodic beacons increase. As a result, efficient operation of ITS suffers due to synchronous beacon collisions. The actual reason for synchronous collisions is the unscheduled channel access mechanism in the IEEE 802.11p [6,7]. In an ad hoc communication setting such as VANETs, the harmonized channel access becomes difficult due to the limited size of the contention window and the aggressive binary exponential back-off (BEB) mechanism. Note that, synchronous beacon collisions can be reduced by reducing the message transmission frequency. However, most of the safety applications have strict frequency requirements [8], therefore, reducing message frequency is not useful for safety applications [9].

It follows that the size of contention window for shared channel access mechanism in IEEE 802.11p must be properly adapted in order to bring time diversity in beacon transmissions by multiple vehicles. We argue that the contention window size adaptation should be based on the underlying channel conditions, given the variation of vehicular density. Moreover, the design should not incur transmission delays due to the increase in the contention window size.

Clearly, the objective of this paper is to provide reliable beacon transmission by minimizing synchronous beacon collisions. In this paper, we propose modifications at the IEEE 802.11p MAC layer that can potentially minimize beacon collisions to improve reliability. A weighted contention window selection is proposed, which replaces the standard BEB in the post transmit phase by using the local channel states. In high density networks, the design also gives prioritized channel access to vehicles experiencing dropped beacons.

The rest of the paper is organized in sections: In Sect. 2, we give necessary background on the IEEE 802.11p standard and presents some observations that lead to the design of the proposed approach. Section 3 describes the proposed weighted contention window adaptation, its behaviour and the algorithm. The evaluation is given in Sect. 3.2. Finally, Sect. 5 concludes the paper.

2   Background

This section gives necessary background on beaconing using the IEEE WAVE networks followed by the transmit power control approaches in the literature.

2.1   The IEEE 802.11p Standard

The IEEE WAVE is a family of standards including, among others: IEEE 1609.1-4 and IEEE 802.11p. The IEEE 802.11p allocates 10 MHz channels each for the Control Channel (CCH) and the Service Channels (SCH) in a 5.9 GHz band for safety and non-safety messages simultaneously. The WAVE devices, i.e. the On-Board Units (OBUs) and the Road Side Units (RSUs), can use both these channel alternatively by switching their radios to a channel defined by the IEEE 1609.4 standard [10]. The time duration to tune a radio to a particular channel is usually set at 50 ms. The CCH is reserved for the safety messages/beacons and it is used simultaneously by all the WAVE-enabled devices. Accordingly, the IEEE 1609.4 standard includes separate functions for different types of messages to be transmitted on the CCH and the SCH.

The most important of these functions is related to the shared channel access mechanism for transmission of beacons on the CCH as shown in Fig. 1. Every transmission is preceded by sensing the CCH. If the CCH is sensed as busy, the transmission is deferred. Otherwise, each transmitting vehicle observes different waiting times before transmission in order to minimize the chances of colliding with other vehicles. The Distributed Inter-frame Space (DIFS) is a time interval, which is observed before attempting to transmit on the CCH. On the other hand, Short Inter-frame Space (SIFS) is representative of a collective time, which includes the time to process a received as well as a response beacon. The beacons are immediately transmitted if the medium is found idle for DIFS duration. If not, the transmitting vehicles select back-off slots from the contention window. Usually, each back-off represents a $13\,\mu s$ slot and it is selected with a uniform random probability from the current contention window. With the passage of every $13\,\mu s$, the back-off decrements by one. When the back-off hits 0, the transmitting vehicle transmits the beacon. If the channel is found busy, then according to Binary Exponential Back-off (BEB) the contention window size is doubled for the next back-off slot selection. Obviously, the probability of synchronous collisions is defined by the size of the contention window.

In the following section, we present some observations about the synchronous collisions in light of the MAC channel access mechanism in IEEE 802.11p standard.

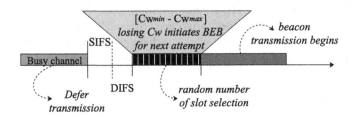

Fig. 1. The mechanism for shared channel access in IEEE 802.11p including the use of contention window and the binary exponential backoff.

2.2 Observations About Synchronous Collisions

Periodic beacons are transmitted using the access category VI as shown in Table 1, which is based on the 802.11e standard [11]. This access category provides a class of service, which has a minimum contention window size of 8 with $cw_{min} = 7$ and $cw_{max} = 15$. The reason for having a small cw_{min} is to transmit beacons before they expire in order to achieve high mutual awareness. Note that, the binary exponential back-off increases the window size upon deferred transmissions and reduces it to the minimum upon a successful transmission. It implies that after a successful transmission, the cw_{min} provides a collision free domain for only 8 vehicles, which causes a high number of synchronous collisions at the start of CCH.

It is also worth mentioning that the BEB was designed to improve the reliability of retransmissions in case of collisions. However, retransmission of beacons in VANETs is not useful due to (1) absence of acknowledgments, and (2) difficulty in judging beacon collisions, which are inherently broadcast in nature. Based on this context, the following observations must be incorporated in the proposed contention window adaptation design to reduce beacon collisions.

Table 1. Contention window sizes defined by the enhanced distributed channel access.

Access category	$cwin_{min}$	$cwin_{max}$
Background	15	1023
Best-effort (AC_{BE})	15	1023
Video (AC_{VI})	7	15
Voice (AC_{VO})	3	7
Legacy DCF	15	1023

Less Aggressive BEB. In IEEE 802.11p, a high-level perspective of a transmission success or failure is indicative of the channel state, that is, a deferred transmission indicates a saturated channel and a subsequent successful transmission indicates a free channel. In VANETs high channel saturation occurs in dense networks and the saturation is likely to persist as long as the vehicle remains a part of the dense network. Therefore, it is safe to say that the channel states are although highly variable in VANETs (defined by the vehicular density), but the change in channel states is not abrupt, as depicted by the aggressive BEB in IEEE 802.11p. Therefore, assuming a constant message frequency, we argue that a contention window adaptation must be less aggressive (i.e. especially after the successful transmission) and adaptive towards channel states, in order to minimize synchronous collisions and to enhance reliable delivery of messages.

Beacon Drops at Source. Another observation originates from the effects of contention window size on the short temporal validity of beacons. That is, the increase in contention window beyond a certain limit increases the probability of dropped beacons at the source, and hence increasing the update delays at the receiver. Also, the exact maximum window size for beaconing is difficult to determine, because contention window adaptation depends upon several dynamic and uncontrollable parameters such as transmission frequency, vehicular density, messages in the queue and channel conditions to name but a few. This notion is significant in adapting the size of contention window up to an extent, which does not affect dropped beacons.

3   The Weighted Contention Window Adaptation Design

Clearly, the weighted contention window adaptation introduces a less aggressive post transmit contention window selection approach by making use of the local information while making sure that increase in the window size does not affect dropped beacons at the source.

To ensure that window adaptation is indicative of the evolving channel conditions (i.e. deteriorating or improving over time) and the contention window adaptation is not aggressive during the post-transmission stage, the design employs two main strategies: (a) a channel congestion state metric to predict the evolving channel condition, and (b) a weighted selection of a suitable post-transmission contention window size for the next beacon.

We use the channel busy time cbt at the physical layer to capture the evolving state of the CCH. According to cbt, the channel is considered busy if the received signal strength is above a certain threshold (i.e. a signal received or collision detected). We record cbt for the previous synchronization intervals (synch-I) i.e. for 10 Hz message frequency, we use 5 synch-intervals. Moreover, the cbt for each synch-I is weighted such that the most recent cbt is weighted higher than the older ones, as follows.

$$cbt(t) = w_1(cbt)_i + w_2(cbt)_{i+1} + \ldots + w_n(cbt)_{i+(n-1)} \tag{1}$$

In order to map the $cbt(t)$ into meaningful weights for the contention window size selection, we introduce a middle contention window size (cw_{mid}) besides the default (cw_{min}) and (cw_{max}) such that $(cw_{min}) < (cw_{mid}) < (cw_{max})$. Then for every successful beacon transmission, the $cbt(t)$ is mapped to a selection probability associated with a contention window size in the post transmit phase as follows:

$$P_{cwin(mid)} = \mid 1 - [\sigma_t * \tau] \mid \tag{2}$$

$$P_{cwin(min)} = 1 - [P_{cwin(mid)}] \tag{3}$$

The $P_{cwin(mid)}$ and $P_{cwin(min)}$ are the probabilities of selecting the middle size contention window and the minimum windows for some value of $cbt(t)$. The σ_t is the inverse of $cbt(t)$ and τ is the threshold of the $cbt(t)$ beyond which weighted contention window selection is considered applicable. As the $cbt(t)$ increases beyond a threshold, the probability of selecting back-off from $cwin_{mid}$ for the next beacon increases. The default IEEE 802.11p BEB is used as long as the channel conditions remain suitable for transmission. That is, upon a successful transmission, the minimum contention window is selected. Moreover, the dropped beacon at the source also forces the proposed approach to select the minimum contention window.

$$cwin_{post-tx} = \begin{cases} cbt > \tau, cwin(mid) \\ cbt < \tau \mid beacondropped, cwin(min) \end{cases} \tag{4}$$

The following section further illustrates the behaviour of the proposed approach.

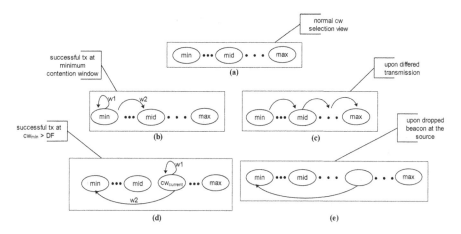

Fig. 2. Behaviour of contention window adaptation during different phases of beacon transmission

3.1 Behaviour at a Microscopic Level

The Fig. 2 illustrates the weighted contention window adaptation mechanism at the MAC layer during different possible stages of beacon transmission: (a) shows the view of the normal contention window with the minimum and maximum window size as defined in IEEE 802.11p standard and the middle contention window size as set by the proposed approach, (b) shows the probability of selecting the minimum window or the middle size upon successful transmission at $cwin_{min}$, defined by the weights $w1$, and $w2$ respectively, (c) shows the increase in window size by $2 * cwin_{current}$ upon a deferred transmission (the increase in window size is similar to the IEEE 802.11p standard), (d) in case of successful transmission at a contention window size, which is higher than the $cwin_{mid}$, the $cwin_{mid}$ is reset to the current window size and then the weights $w1$ and $w2$ are applicable as in Fig. 2(a), finally in (e) upon dropped beacon at the source, the window size is set to the minimum window size with the probability 1.

Note that, the selection of back-off from $cwin(mid)$ for a subsequent beacon after successful transmission has implications on dropped beacons at the source. That is, continuous transmissions at a higher contention window may result in longer waiting times in the queue and, as a result, dropped beacons before transmission. Under such conditions, as soon as a vehicle detects a dropped beacon, the back-off is immediately initialized to $cwin_{min}$ to reconcile for the delay incurred due to the loss of the dropped beacon.

For the sake of logical argument and to highlight the usefulness of the proposed approach, we consider the following example:

Without loss of generality, let's assume that two vehicle v_i and v_j have similar values for cbt, then the probability of simultaneous transmission by selecting same back-off is given by $P(v_i = v_j)$. Where $v_i = s$ for $s \in [all\ slots\ in\ cw_{min} \bigwedge cw_{mid}]$ containing initial and maximum contention windows sizes of c_{min} and c_{mid} respectively, then selecting s_i and s_j by v_i and v_j respectively are independent events. So, we have Eq. 5.

$$P(v_i = v_j) = \sum_{x=c_{def}}^{c_{mid}} P(v_i = s \mid v_j = s) \tag{5}$$

Since, $P(v_i = s) = P(v_j = s)$ for every slot in the contention window, therefore it is sufficient to calculate the $P(v_i = s)$. Hence, for $s \in [cw_{min} : cw_{mid}]$, we have the law of total probability:

$$P(v_i = s) = P(v_i = s \mid cw_{min}).P(cw_{min}) + P(v_i = s \mid cw_{min})$$

$$.P(cw_{mid}) = \begin{cases} \frac{1}{|cw_{min}|}.w_{cw_{mid}} + \frac{1}{|cw_{mid}|}.w_{def}, & s \in cw_{min} \\ 0.w_{cw_{mid}} + \frac{1}{|cw_{mid}|}.cw_{mid}, & s > cw_{min} \end{cases} \tag{6}$$

Thus, the probability of synchronous collision due to same back-off selection between two vehicles $P(v_i = v_j)$ with same $cwin_{min}$ and $cwin_{mid}$, is given by:

$$P(v_i = s) = \sum_x P(v_i = x, v_j = x) = \sum_a P(v_i = x)^2 \tag{7}$$

The benefit offered by the weighted contention window selection is the probabilistic post-transmission selection of $cwin_{min}$, which is a less aggressive approach and minimizes collisions at the start of CCH. In addition, vehicles experiencing high slot utilization can also select back-off from $cwin_{min}$ with certain reduced probability. It means that high slot utilization does not always allocate a large window size and presents an opportunity for vehicles to transmit using small window size. In addition, to avoid vehicles from continuous transmissions using a higher window size, the proposed approach uses a dropped beacon as an indication for very long waiting times at the source. Therefore, to provide prioritized channel access to account for the dropped beacon, the window size is initialized to $cwin_{min}$ for the next beacon transmission.

3.2   Algorithm: Contention Window Adaptation

The algorithm for contention window adaptation is given in Algorithm 1. The inputs to this algorithm are the beacons from the application layer, transmission status and the value of cbt. The algorithm gives the probabilities for selecting a contention window size upon each transmission attempt $(cwin_{(post-tx)})$. Initially, the algorithm demarcates the contention window sizes i.e. $cwin_{min}$, $cwin_{mid}$ and $cwin_{max}$ in line 1. Then the back-off for all beacons arriving from the application layer is selected using the function Backoff() at line 3. The arguments of this function are $P_{(cwin(mid)}$ and $P_{(cwin(min))}$, which specify the probability of selecting a post transmit back-off from $cwin_{mid}$ and from $cwin_{min}$, respectively. The line 5 through line 7 records the cbt during the back-off interval and in line 8 the beacon is transmitted. The algorithm from line 11 through line 25 is significant in order to record the transmission status and to convert the slot utilization into

---

Algorithm 1. Contention Window Adaptation
inputs: *beacons, transmission status, cbt(t)*
outputs: $cwin_{post-tx}$

1: set $(cwin_{mid}) \mid (cwin_{min}) < (cwin_{mid}) < (cwin_{max})$
2: **for** beacons from above **do**
3:     **procedure** BACKOFF $(P_{cwin(mid)}, P_{cwin(min)})$
4:         pick $backoff \leftarrow [cwin_{min} - cwin_{mid}]$
5:         **while** $backoff$ **do**
6:             record $cbt(t) \leftarrow equation\ 1$
7:         **end while**
8:         transmit
9:     **end procedure**
10: **end for**
11: **if** $(cwin_{current} > cwin_{mid})$ **then**
12:     $cwin_{mid} = cwin_{current}$
13: **end if**
14: **switch** *transmit status* **do**
15:     **case** *transmitted*
16:         calculate $cwin_{post-tx} \leftarrow equation\ 4$
17:         call Backoff()
18:     **case** *deferred*
19:         set $cwin_{current} \leftarrow ((cwin_{current}(v_i) + 1) * 2) - 1$
20:         calculate $cwin_{post-tx} \leftarrow equation\ 4$
21:         call Backof()
22:     **case** *Dropped*
23:         set $P_{cwin(min)} = 1$
24:         $P_{cwin(mid)} = 0$
25:         call Backoff()

---

meaningful weights that can be used to determine the contention window size for the next beacon transmission. First of all at line 11, the current contention window size is checked and if it is greater than the $cwin_{mid}$, then the $cwin_{mid}$ is reset to $cwin_{current}$, otherwise, the contention window size demarcation remains the same as in line 1. The transmission at line 8 may result in a successful transmission, a deferred transmission or a dropped beacon during the back-off. As such for a successful transmission, the $cwin_{(post-tx)}$ is calculated using Eq. 4. For deferred transmission, the contention window is increased as specified in IEEE 802.11p and then $cwin_{(post-tx)}$ is calculated. In either case, the calculated values for $P_{(cwin(min)}$ and $P_{(cwin(mid)}$ are used to call the Backoff() function at line 17 and line 21. Finally, if the beacon is dropped during the back-off, the value of $P_{(cwin(min))}$ is set to 1 and $P_{(cwin(min))}$ is set to 0. It indicates that for the next beacon transmission the back-off at line 4, will be selected from the $cwin_{min}$. This shows the prioritized channel access mechanism to make up for the previous dropped beacon.

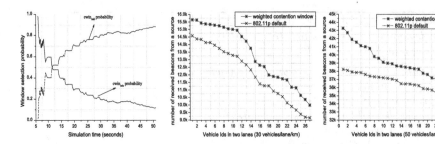

Fig. 3. Run-time selection probability of minimum contention window w.r.t CBT for the first few seconds of simulation

Fig. 4. Awareness quality measured as the number of received beacons in 50 veh/lane/km scenario

Fig. 5. Awareness quality measured as the number of received beacons in 30 veh/lane/km scenario

This concludes the specification of the weighted contention window adaptation approach which aims to reduce overall synchronous collisions in the network. In the next section, we evaluate the proposed approach.

4 Evaluation of Contention Window Adaptation

This section evaluates the weighted contention window approach proposed in this paper. First, we verify the correct functioning of the proposed design followed by a comparison with the de facto standard i.e. IEEE 802.11p.

The Veins framework – version 2.1, OMNeT++ – version 4.2.2 and sumo – version 0.17.0 is used for evaluation. The WAVE application layer is configured to generate beacons at 10 Hz. The MAC layer is responsible for acquiring channel states from the physical layer. The simulation scenario consists of the 1 Km 2 way and 4 way highways with varying number of vehicular densities freeway speeds.

4.1 Results

As aforementioned, when a vehicle transmits a beacon, the proposed approach monitors the channel states in order to associate a meaningful weight for contention window size selection. Therefore, the implementation of weighted contention window requires modifications at the MAC layer during the post transmit phase.

The logic behind weighted contention window is to associate probabilities with minimum and middle contention window sizes with respect to the increasing channel saturation. Therefore, it is important to verify this behaviour for vehicles in a simulated scenario. We configure a two lane highway which is heavily populated with vehicles that transmit beacons at a high frequency. In Fig. 3, for increasing vehicular densities, we record the window selection probabilities for minimum and middle window sizes in the post transmit phase. It could be observed that as the channel becomes saturated (here increase in time is representative of the increasing number of vehicles or otherwise more congestion), the probability of middle contention window approaches to 1. Accordingly, the with the exact same proportions, the minimum window selection probability approaches to 0. This behaviour verifies the evolution of weights for window sizes according to the design.

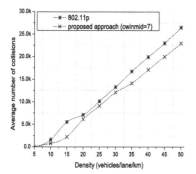 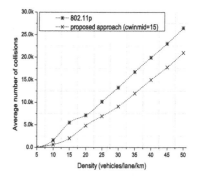

Fig. 6. Comparison of average number of collisions for varying levels of vehicular densities and window sizes with $cwin_{mid}$ set at 7

Fig. 7. Comparison of average number of collisions for varying levels of vehicular densities and window sizes with $cwin_{mid}$ set at 15

One way of measuring awareness is to measure the number of received beacons in a network. Clearly, high message reception means a high level of awareness of the local topology. In Figs. 4 and 5, the number of received beacons from a source vehicle is recorded on different vehicles. The receiving vehicles are arranged on x-axis with respect to their increasing distances from the source. By controlling the synchronous collisions, the awareness quality in terms of the proposed approach increases as compared with the IEEE 802.11p.

High message reception is achieved due to the less aggressive behaviour in selecting the $cwin_{min}$ and larger window sizes in the post transmit phase. The Figs. 6 and 7 shows the average number of collisions. Observe that, significantly fewer collisions are recorded for the proposed approach as compared with the IEEE 802.11p. Besides, for higher values of $cwin_{mid}$, the collisions are further reduced.

In a highway scenario of 50 vehicles/lane/km in a two lane road, we show the performance of the proposed approach using overall throughput. In Fig. 8, the results are compared with the standard IEEE 802.11p. It can be observed that initially for few seconds the throughput values remain similar. This is because initially the network has limited vehicles and the probability of selecting the minimum contention window remains very high. However, as the number of vehicles increase, the proposed approach starts to select $cwin_{mid}$ in the post-transmit phase for new beacons. Therefore, as a result of reduced collisions, a higher throughput can be observed.

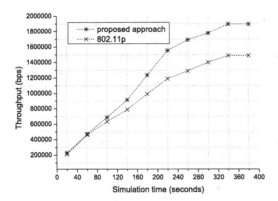

Fig. 8. Comparison of throughput variation of the proposed approach with the standard 802.11p

5 Conclusion

The stipulated amendments in the WAVE offer little relief to the problem of synchronous collisions. In this paper, we identified the limitations of the contention window size and the aggressive BEB as main reasons for synchronous collisions. The proposed contention window adaptation approach is proposed, which translates the channel busy times into meaningful weights for selecting the window size in the post transmit phase. After a successful transmission, the default aggressive behaviour of BEB is replaced such that a higher probability of selecting the minimum window is applicable in situations of less channel saturation and vice verse. Moreover, the window adaptation design also makes provisions for prioritized channel access to vehicles experiencing dropped beacons. The simulation results clearly demonstrates reliable beacon transmission as compared to the IEEE 802.11p standard.

Acknowledgment. The authors would like to thank the High Impact Research (HIR), University of Malaya for the support and funding.

References

1. Hartenstein, H., Laberteaux, K.P.: A tutorial survey on vehicular ad hoc networks. IEEE Commun. Mag. **46**(6), 164–171 (2008)
2. Shah, S.A.A., Shiraz, M., Nasir, M.K., Noor, R.B.M.: Unicast routing protocols for urban vehicular networks: review, taxonomy, and open research issues. J. Zhejiang Univ. Sci. C (Comput. & Electron.) 15(7), 489–513 (2014). http://www.zju.edu. cn/jzus/iparticle.php?doi=10.1631/jzus.C1300332
3. Ahmad, I., Ashraf, U., Ghafoor, A.: A comparative QoS survey of mobile ad hoc network routing protocols. J. Chin. Inst. Eng. **39**(5), 585–592 (2016). http://dx.doi.org/10.1080/02533839.2016.1146088

4. S.A.A. Shah, E. Ahmed, F. Xia, A. Karim, M. Shiraz, R.M. Noor: Adaptive bea-coning approaches for vehicular ad hoc networks: a survey. IEEE Syst. J. **PP**(99), 1–15 (2016). http://ieeexplore.ieee.org/xpl/articleDetails.jsp?arnumber=7506244
5. Uzcategui, R., Acosta-Marum, G.: Wave: a tutorial. IEEE Commun. Mag. **47**(5), 126–133 (2009)
6. Stanica, R., Chaput, E., Beylot, A.-L.: Loss reasons in safety vanets and impli-cations on congestion control. In: Proceedings of the 9th ACM Symposium on Performance Evaluation of Wireless Ad Hoc, Sensor, and Ubiquitous Networks, pp. 1–8. ACM, New York (2012)
7. Eckhoff, D., Sofra, N., German, R.: A performance study of cooperative awareness in ETSI ITS G5 and IEEE WAVE. In: 10th Annual Conference on Wireless On-demand Network Systems and Services (WONS), pp. 196–200. IEEE (2013)
8. C.V.S.C. Consortium, et al.: Vehicle safety communications project: task 3 final report: identify intelligent vehicle safety applications enabled by DSRC. National Highway Traffic Safety Administration, US Department of Transportation, Washington DC (2005)
9. Park, Y., Kim, H.: Collision control of periodic safety messages with strict mes-saging frequency requirements. IEEE Trans. Veh. Technol. **62**(2), 843–852 (2013)
10. IEEE standard for wireless access in vehicular environments (wave)-multi-channel operation, IEEE Std 1609.4-2010 (Revision of IEEE Std 1609.4-2006), pp. 1–89, February 2011
11. Mangold, S., Choi, S., May, P., Klein, O., Hiertz, G., Stibor, L.: IEEE 802.11 e wireless lan for quality of service. In: Proceedings of European Wireless, vol. 2, pp. 32–39 (2002)

Model Driven Architecture for Decentralized Software Defined VANETs

Afza Kazmi[1]([⊠]), Muazzam A. Khan[1], Faisal Bashir[2], Nazar A. Saqib[1],
Muhammad Alam[3], and Masoom Alam[4]

[1] NUST College of EME, National University of Sciences and Technology (NUST),
Islamabad, Pakistan
afzakazmi@gmail.com, {muazzamak,nazar.abbas}@ce.ceme.edu.pk
[2] Department of Computer Science, Bahria University, Islamabad, Pakistan
faisalwn@yahoo.com
[3] Instituto de Telecomunicaciones, University of Aveiro, Águeda, Portugal
[4] Department of Computer Science, COMSATS Institute of Information Technology,
Islamabad, Pakistan

Abstract. Vehicular Adhoc Networks (VANETs) are considered as
a breakthrough technology in providing robust Vehicular communication that guarantees Road safety and offer interactive set of applications. VANET faces certain orchestration and management challenges.
Recently, the notion of Centralization of VANET is gaining importance. In this regard SDN (Software defined Network) is integrated with
VANET to achieve management goals. SDN renders existing VANET
architecture to provide centralized control. Yet for Large-scale VANET,
SDN approach does not outperform due to SPOF (single point of failure) in SDN. We have proposed decentralized architectural solution that
scales out overall network intelligence into respective local controllers
that result in unprecedented elasticity and resource availability. To make
this approach operational no standard Architectural approach yet exists.
An emerging Model Driven Architecture approach is used that simplifies
VANET development using abstract models. It only subdues VANET
rigidity but also comply with SDN principals. This system is implemented in VEINS framework. Results demonstrate its efficacy for large
scale VANET.

Keywords: Vehicular adhoc network · Software defined network ·
Model driven architecture · VEINS · Highway

1 Introduction

In the recent era of technological uplift, vehicular networking is considered a
promising and enabling technology that helps in realization of a diversification of
vehicle related applications. Intelligent Transportation System (ITS) streamline
such Networks and makes it viable for road traffic safety, emergency management, and infotainment [1]. Self-organization of vehicles in conventional VANET

© ICST Institute for Computer Sciences, Social Informatics and Telecommunications Engineering 2017
J. Ferreira and M. Alam (Eds.): Future 5V 2016, LNICST 185, pp. 46–56, 2017.
DOI: 10.1007/978-3-319-51207-5_5

confronts several Management challenges such as dynamic network topology, high mobility and unbalanced vehicular traffic. Apart from VANET communication challenges, standard VANET design and architectural model is always overlooked. Lack of centralization of control in VANETs has always remained an intractable issue. Centralization of control enables the network manager to view the network in global context. The concept of centralized VANET is realized by emerging SDN technology [7]. SDN segregates network intelligence from infrastructure. SDN adds flexibility in network through programming, which allows the network stakeholders to adapt VANET according to their needs. Existing VANET architectures provide state-of- the-art approaches to build proficient VANETs, but when network size increases, it becomes difficult to manage the extensive data plane requests. One of such significant factor is Scalability in SD-VANET (Software defined VANET). The term textitScalability is used to describe vehicle density for large road networks [6]. Secondly, there is absence of standard VANET architecture. Both experts handle different level of abstractions that clearly signifies a communication gap between them. Ultimately, this gap causes incomplete and inconsistent requirement engineering. These concerns are better addressed in MDA approach in which the system is broken down into its abstraction levels [9]. The organization of the paper is summarized as; Sect. 2 discusses state-of-the-art VANET architectures and provide critical analysis. Section 3 include design requirements and operational mode of distributed control plane structure of DSDVANET (Decentralized software defined VANET) and building models based on these requirements, a comprehensive process of MDA development in DSDVANET are discussed. Section 4 provides the simulation environment where proposed model is deployed. Results are made by taking parameters from existing VANET models. Section 5 concludes the research by validating the existing model and its scope for future VANET systems.

2 Literature Work

Initial efforts reveal the advent of various wireless communication standards in VANET [2]. Different VANET protocol standards and architectures were developed in European countries. US DOT [10] are taken as initial standardization effort in ITS. A top-down Information centric architecture is proposed [11] which provides information enriched VANET applications, using the design concept of three key features space, time and users. An Integration of Cloud Computing and Information Centric Networking has been used that deploy mobile cloud model to VANET, and optimizes data routing and dissemination [12]. Similarly WiMAX and DSRC capabilities are combined in VANET to provide internet access to the vehicles in [13]. Cellular network technologies are also leveraging VANETs capabilities. LTE4V2X, offers centralized VANET schemes using LTE [3]. In [7] SDN is adapted for VANET environments. In this architecture OpenFlow enabled switch is contained inside vehicle. This architecture provides different operational modes which are based on degree of control of the Controller. In [14] SDN and Fog Computing are integrated that provides location oriented and delay-sensitive services for next generation VANETs. A Vehicular Cloud [15] consists of traditional and specialized RSUs, using SDN it helps network in handling multiple data plane requests and minimizes control plane (CP) overhead. A novel IEA (information exchange architecture) [16] handles several data e.g. the road traffic flows, vehicle tracking, based on the data forwarding scheme. In [5] characteristics of the inter-contact time and duration of contacts for vehicular networks is determined for SDN based VANET. A centralized scheduler is installed at RSU by using SDN concepts. In MDSE paradigm a nominal research work has been done in developing WSN applications [17]. In [4] MDA principals are used to construct wireless sensor and actuator application, here MDA partitions network into respective abstraction layers, and each layer is developed using specific DSML (Domain specific modeling language). We evaluate these architectures on the basis of selective parameters to assess their performance using Network QoS as well as Software Quality Attributes. Combined WiMAX/DSRC VANET and SDN based VANET have inherent. From modeling perspective most of the Vanet system are built using networking models only two VANET architectures support software oriented models namely InVanet and combined WiMAX/DSRC. Combined WiMax/DSRC based VANET can perform model/code transformations. LTE4V2X and FSDN VANET provides services for large-evolvable VANETs. From this analysis we deduced that the aspect of software oriented models are less applied. SDN provides the opportunity to leverage MDSE (Model driven Software Engineering) in VANET context. It enables in constructing flexible and interpretable models for building Scalable VANET. A Large-scale VANET system comprises of multiple domains (cities or regions). A major bottleneck of such network is Scalability, where a single SDN controller has to manage multiple domains that results in substantial performance degradation [18]. We have introduced an approach to scale-out the intelligence of such network where each domain has its own local intelligence. Next section enlightens MDSE role in SDN and VANET domains.

3   Proposed Methodology

Recent research on distributed SDN is conducted in Data centers where network is scaled out to control billions of data plane requests from various sites across the globe [22]. Our proposed architecture not only believes on fully decentralized intelligence but carefully managing distributed controllers by top RC (Root Controller) that have a global network view and supremacy to perform centralized management. The proposed model is novel from different aspects.

- In existing SDVANETs the RSU are solely assigned to perform Forwarding tasks. In our proposed architecture RSUs acts as OpenFlow-enabled switches to implement forwarding rules enforced by the controller.
- MDA principles are combined with SDVANET that add abstraction and flexibility to the network.
- The CP (Control Plane) distribution is made in hierarchical fashion that delegates the roles from RC to Domain/Local controller(s) (Fig. 1).

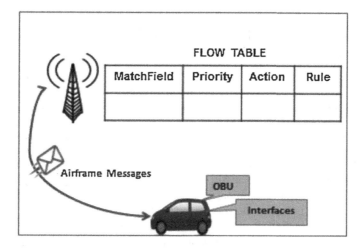

Fig. 1. Generic structure of forwarding plane.

3.1 Design Requirements

Due to distant positions of both Controller and RSUs there is high probability of delayed response. For safety related VANET applications time of response is the key constraint. Hence our design approach is based on Scalability requirement. We are going to exploit the middle layer that is the CP of the model. Our aim is to distribute the centralize intelligence into respective local territories termed as domains. We are examining our proposed approach on a Highway, connecting several regions (Domains). Each domain has its own centralized controller called DC (domain controller). The functional distribution is made in such a way that RC performs the most specialized functions while the generalized functions are handled by DCs.

DSDVANET Forwarding Plane Requirements: The core element of this layer is RSU that acts as an intermediary between moving vehicles and DC. RSU is connected via WAVE interface with the vehicles to provide fast communication in a single domain. It acts as Wireless OpenFlow-enabled switch (for Data Plane Communication) as well as Ethernet switches (for CP Communication). A primary Design requirement of Forwarding/Data plane is to enable Scalability that allows multiple distant nodes and RSUs to communicate with each other inside a domain. RSU maintains the Flow entries in the Flow Table (FT). The structure of OpenFlow enabled RSU is given below. Each flow entry of FT has some properties such as priority, MatchField, Action, Counters, timeouts etc. When a vehicle sends packet to the RSU it traverses the Flow Table to lookup for the Match of the respective flow entry [7]. If a match takes place the packet is processed otherwise the packet is send to CP that decides respective actions.

DSDVANET Control Plane Requirements: The CP of the Decentralized SDVANET is extensible as well as the crucial part to design. We partition this plane, in order to achieve network wide orchestration goals. CP-Layer1 holds RC and the subsequent layer holds DC. From the concepts of distributed computing [8] the CP can be partitioned using two approaches:-

Vertical Partitioning: In this approach there exists a single physical machine (controller), which is distributed hierarchically [8]. The tasks of the CP are assigned to different controllers. The lower level controllers(DCs) perform frequent events' taking place in Data Layer while the upper layer handle global events.

Horizontal Partitioning: It is a physical distribution setting in which CP is divided into domains. Each domain has a DC which is responsible for its own territory. DC manages a disjoint set of *OpenFlow-Enabled RSUs* (Switches). DCs communicate with its adjacent to enforce global policies [19]. In our proposed model the distributed CP takes advantage of both distribution techniquesby relating their strengths. CP is partitioned into two layers, both hierarchically and physically. The top layer holds RC. The successive layer holds DCs, which supervises their own domain/territories. Each DC is connected to RC via dedicated high speed broadband connections. DC access RC, when unexpected events occur. e.g. when DC fails, network state changes, faulty links/flow tables discovered or DC receives out of domain information.

3.2   Operational Mode of Control Plane

Control Plane: Layer-1: It contains most specialized modules e.g. Flow Rules Manager and SDN Repository. DC acknowledges RC about its domain level activates by sending information such as link State, Vehicle/Nodes Availability, DC status etc. All this information is gathered for a scheduled time period inside SDN repository. Moreover it also keeps the network topology patterns as they kept on changing in VANET. In Forwarding Engine RC configures its submissive controllers (DC) based on Forwarding rules. These rules are developed from Domain level information. Path Computation Module provides a mathematical model to generate multi path topology based on efficient VANET routing schemes. Error Handling troubleshoots SDVANET issues and respond quickly by providing an optimal solution. The typical Errors handled by this module are Broken Links, Connectivity deadlocks etc. [21].

Control Plane: Layer-2: This is an outcome of horizontal scaling of CP. It contains physically distributed DCs. DC holds their respective territories and also connected with neighboring DCs to maintain consistent network topology. DC performs the similar tasks of centralized SDN Controller in SDVANET. The main functionalities of DC are; Domain Network Services and Domain Request Handling. *Domain Network Services*: These services are delegated to the DP elements to augment the network performance by reducing traditional VANET bottlenecks such as dynamic mobility, connectivity loss and Flow Management. DC serves the frequent events coming from Data plane. The main focus of DC is to provide high responsiveness and scalable deployment of network devices (RSUs).

Domain Request Handling: handles the requests coming from Data plane (DP). The *Dispatcher* module captures all the DP requests and processes them accordingly. It serves as the entry point to the CP. It gathers and forwards the DP requests to the *Configurator*. *Configurator* is the local agent which acts as a CPU of the DC. It is a visualization tool that displays the DC load statistics and provides a summary of Dispatcher-Configurator sessions. It is helpful in estimating Network performance at domain level. It timely collects the information from the configurator and identifies the condition of configurator overload (Fig. 2).

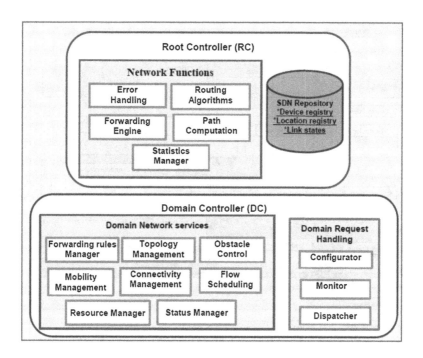

Fig. 2. Distributed control plane of Decentralized SDVANET

3.3   Mapping Network Requirements to Models

Preliminary activity of MDA is analyzing the requirements gathered from stakeholders. *Scalability* is Mission Statement of this proposed methodology. Next targeted users are identified. Domain experts are the application developers that develop application of SDVANET. *Network experts* perform administrative tasks e.g. topology configuration, protocol selection, and forwarding rules enforcement. Communication with CP is possible through an interface layer that holds protocols such as *OpenFlow* [23]. *PIM Meta Models:* is a visual representation of structural and behavioral details of Decentralized SDVANET modules. A generic DSML is used (UML) for describing internal and external system behaviors. Logical view of DSDVANET: Model diagram is used for modeling this SDN layered system. The stereotypes inside each model (data, control, application) show the interactions and dependencies between the components. Application Plane model holds Applications package that handles generic application functionalities which are used in derived application packages in RC (as*SimpleServerApp*), and DCs (as *CtrlApps*), RSU (as *WAVEApp*)and vehicle (*TraCI*) applications (Fig. 3).

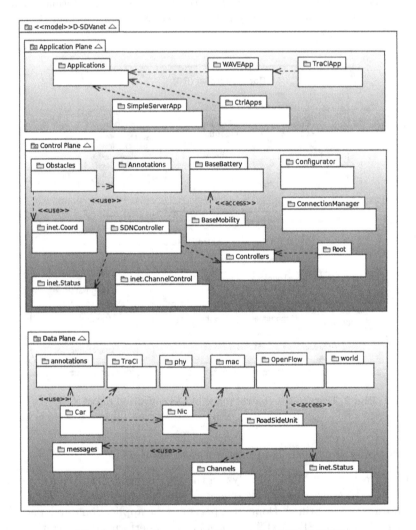

Fig. 3. Model diagram of Decentralized SDVANET

3.4   Defining Classes and Hierarchy

Due to extensive set of network modules, we have covered most significant artifacts. *Coord* class stores the dimensions of each object (vehicle, RSU or obstacle) in an ITS Network. It plays a vital role in generating mobility patterns and computing distance in V2V and V2I for effective communication. *CtrlBehavior* is an interface which implements various controller modes of operations. Controller mode of operations are *Hub*, *Forwarding* and *Switch*. Hub class does not contain any intelligence, it instructs the RSU to flood all packets on all ports. *Switch* class learns the mapping of network elements to make easy re-configuration of RSUs if needed. *Forwarding* class contains prior knowledge of the whole network. *AnnotationManager* is composed of several geometrical classes (Line, Polygon) used to design virtual network for simulation. It manages the annotations on OMNET++ canvas. *Transformation Rules* ensure consistency between models at PIM with PSM. They are designed by considering the structural details of the system. In DSDVANET it is difficult to implement the transformation rules as we have to keep SDN policies consistent with MDA principals.

- *Level 1 (Network level)*: holds Network wide models which describes the overall data flow. These models are independent of Network. From PIM models of D-SDVANET such models are Car, RoadsideUnit, DC and RC.
- *Level 2 (Domain Level)*: are the domain VANETs. They depend on Network. It include SDN elements and Flow Processing models to enforce SDN rules.
- *Level 3 (Node Level)*: It is the lowest level and provides behavioral models of each vehicle. It holds the models that describe the contextual details of node. e.g. Channel Selection, Communication links, Protocol Selection and Interface Modeling. The transformation rule maps the similar elements of the Network and Domain-level models. It gives a fixed default value to the elements that newly appear in the Domain-level model. Same Step is repeated while transforming Domain-level to Node-Level. After developing PSM Models, we export these models to VEINS simulator [22], combined with OpenFlow Extension [23], to provide SDN facility in VANET.

4   Results and Discussion

After developing PIM Models, we exported the models to respective PSM tool (i.e. VEINS). In our scenario two cities are connected Peshawar and Islamabad. The network topology is partitioned into domains. Each domain holds a ping application that randomly send echo message to any of the 9 domains. On receiving, echo reply message calculates mean RTT. The total simulation time for this experiment is 125 s. We have utilized the following metrics to analyze the system performance.

Mean RTT (Round Trip Time): When a domain sends an echo request message to its linked domain and receive acknowledgment in form of echo reply message the time elapsed is taken as RTT. The simulation duration of each run is 30 s. We calculate RTT for 10 runs per domain and convert it into seconds. Mean RTT is described with respect to RC location indices. It can be seen that controller placement at some equidistant location like 4 and 5 provides best results in the form of least mean RTT for each domain of the network topology. However

moving towards the edge locations give worst results, that is high RTT. Green
and Blue bars shows steady rise in RTT if RC is placed on the other extreme
end of the network topology however red bars depict a stability of RTT among
all domains (Fig. 4).

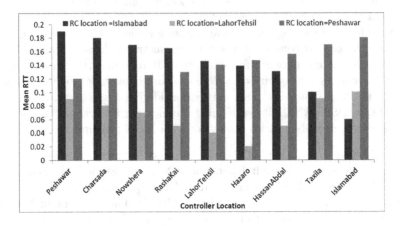

Fig. 4. Mean RTT Vs. RC location. (Color figure online)

Gain of Scalability (GoS): For a simulation time interval T, the number of Vehicles served either via V2I/V2V communication is termed as GoS [4]. We have
defined 5 different traffic scenarios in Table 1 for 3 lanes (L1, L2 and L3) of Highway road. The sender node receives 64 RTTs per domain as each sender node
waits for 2 s to get a reply and after that sends echo to another domain. From
Fig. 5, in Traffic Scenario 1 node density is high as compared to the other roads.
Certainly a reduction in vehicle speeds is observed. However in SDVANET and
DSDVANET cases we observe a slight drop, due to handover between RSU and
Controller. In 2nd and 3rd Scenario VANET faces a descent due to change in
vehicle speed and road length. However SDVANET and DSDVANET maintained
their levels as Topology management in controller keeps the network topology
updated.

Table 1. Different traffic scenarios for M1 motorway

Sr	Distance source↔ Destination (km)	Mean vehicle speed (km/h)			Mean vehicle arrival rate (vehicle/h)			Mean vehicle density (vehicles/km)
		Lane1	Lane2	Lane3	Lane1	Lane2	Lane3	
1	Peshawar↔Charsada(32)	104	95	36	1163	38	17	24.3
2	Hazro↔Taxila(45)	95.5	96	50	554	42	29	20.6
3	Taxila↔Islamabad(40)	86.8	98	52	898	46	25	21.5
4	Charsada↔RashaKai(31)	70.59	100	62	740	24	20	16.3
5	LahorTehsil↔Hazro(63)	65	100	38	537	35	26	19.7

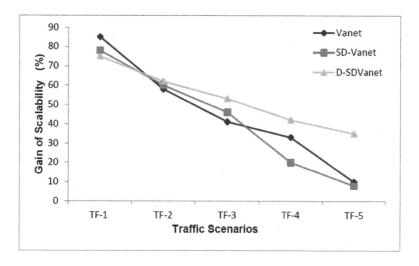

Fig. 5. GoS vs Traffic scenarios

5   Conclusion and Future Work

In this research concept of DSDVANET is introduced that reduces scalability concerns. Practicing MDA approach in VANET domain is found to be useful as it enable flexible and consistent model development. MDA and SDN both exhibit several similar features that help in generating consistent architectural models. MDA extensibility feature helps is building standard VANET architecture. This scheme is designed for large VANET systems, for limited VANET SDVANET is suitable. It is validated in real-world Motorway scenario. Results has been shown that the controller perform effective role in handling DP requests and RC resolves the issue of VANET failure. In future we will extend application plane of DSDVANET to deploy non-safety VANET applications.

References

1. Alam, M., Ferreira, J., Fonseca, J.: Introduction to intelligent transportation systems. In: Alam, M., Ferreira, J., Fonseca, J. (eds.) Intelligent Transportation Systems. SSDC, vol. 52, pp. 1–17. Springer, Heidelberg (2016). doi:10.1007/978-3-319-28183-4_1
2. Campolo, C., Molinaro, A., Scopigno, R.: From today VANETs to tomorrow planning and the bets for the day after. Veh. Commun. **2**, 158–171 (2015)
3. Remy, G., Senouci, S.-M., Jan, F., Gourhant, Y.: LTE4V2X: LTE for a centralized VANET organization. In: IEEE GLOBECOM (2011)
4. Rodrigues, T., Batista, T., Delicato, F.C., Pires, P.F., Zomaya, A.Y.: Model-driven approach for building efficient wireless sensor and actuator network applications. In: SESENA, USA (2013)
5. Xiao, X., Kui, X.: The characterizes of communication contacts between vehicles and intersections for software-defined vehicular networks. Mob. Netw. Appl. **20**(1), 98–104 (2015)

6. Kazmi, A., Khattak, M.A., Akram, M.U.: DeVANET: decentralized software-defined VANET architecture. In: 3rd IEEE Symposium (SDS-2016) (2016)
7. Ian, K., You, L., Gerla, M., Gomes, R.L., Ongaro, F., Cerqueira, E.: Towards software-defined VANET: architecture and services. In: 13th IEEE Annual Mediterranean Ad Hoc Networking Workshop (MED-HOC-NET) (2014)
8. Bhowmik, S., Tariq, M.A., Koldehofe, B., Kutzleb, A., Rothermel, K.: Distributed control plane for software-defined networks. In: 9th ACM Conference on Distributed Event-Based Systems (2011)
9. Araniti, G., Campolo, C., Condoluci, M., Iera, A., Molinaro, A.: LTE for vehicular networking: a survey. IEEE Commun. Mag. **51**, 148–157 (2013)
10. TERIS Website, Official web site of the (National ITS Architecture)
11. Bai, F., Krishnamachari, B.: Exploiting the wisdom of the crowd: localized, distributed information-centric VANETs. IEEE Commun. Mag. **48**, 138–146 (2010)
12. Lee, E., Lee, E.-K., Gerla, M., Oh, S.Y.: Vehicular cloud networking: architecture and design principles. IEEE Commun. Mag. **52**, 148–155 (2011)
13. Alam, M., Albano, M., Radwan, A., Rodriguez, J.: CANDi: context-aware node discovery for short-range cooperation. Trans. Emerg. Tel. Tech. **26**(5), 861–875 (2013)
14. Truong, N.B., Lee, G.M., Ghamri-Doudane, Y.: Software defined networking-based vehicular adhoc network with Fog computing. In: IFIP/IEEE International Symposium on Integrated Network Management (IM) (2015)
15. Salahuddin, M.A., Al-Fuqaha, A., Guizani, M.: Software-defined networking for RSU clouds in support of the internet of vehicles. IEEE Internet Things J. **2**(2), 133–144 (2015)
16. Ryoo, I., Na, W., Kim, S.: Information exchange architecture based on software defined networking for cooperative intelligent transportation systems. J. Cluster Comput. **18**(2), 771–782 (2015)
17. Rodrigues, T., Dantas, P., Delicato, F.C., Pires, P.F., Pirmez, L., Batista, T., Miceli, C., Zomaya, A.: Model-driven development of wireless sensor network applications. In: 9th IEEE/IFIP Conference on Embedded and Ubiquitous Computing (2011)
18. van Asten, B.J., van Adrichem, N.L.M., Kuipers, F.A.: Scalability and resilience of software-defined networking: an overview (2014)
19. Schmid, S., Suomela, J.: Exploiting locality in distributed SDN control. In: HotSDN 2013, 2nd ACM SIGCOMM Workshop on Hot Topics in Software Defined Networking
20. Yeganeh, S.H., Ganjali, Y.: Kandoo: a framework for efficient and scalable offloading of control applications. In: (HotSDN), pp. 19–24 (2012)
21. Phemius, K., Bouet, M., Leguay J.: DISCO: distributed multi-domain SDN controllers. In: Network Operations and Management Symposium (NOMS) (2014)
22. VEINS. http://veins.car2x.org/documentation/
23. Klein, D., Jarschel, M., An OpenFlow extension for the OMNeT++ INET framework. In: Asia-Pacific Conference on Computer Aided System (2014)

Data Rate Adaptation Strategy
to Avoid Packet Loss in MANETs

Muhammad Saleem Khan[1]([✉]), Mohammad Hossein Anisi[2],
Saira Waris[1], Ihsan Ali[2], and Majid I. Khan[1]

[1] Department of Computer Science,
COMSATS Institute of Information Technology, Islamabad, Pakistan
skhan.ciit@gmail.com, syra054@yahoo.com, majid_iqbal@comsats.edu.pk
[2] Faculty of Computer Science and Information Technology,
University of Malaya, Kuala Lumpur, Malaysia
anisi@um.edu.my, ihsanalichd@siswa.um.edu.my

Abstract. Due to mobile and dynamic nature of MANETs, congestion
avoidance and control is a challenging issue. Congestion mainly occurs
due to the phenomena where data arrival rate is higher than the transmission rate of data packets at a particular node. Congestion results in high
packet drop ratio, increased delays and wastage of network resources. In
this paper, we propose data rate adaptation technique to avoid packet
loss. Proposed technique is based on the analysis of queue length of the
forwarding nodes, number of source nodes forwarding data through a
particular forwarding node, and rate of link changes. In the proposed
strategy, queue length of forwarding nodes is communicated periodically
to the neighbor nodes. Keeping in view the queue length of forwarding node, the sending node adapts its sending data rate to avoid congestion and to ensure reliable data communication. Results show that
proposed strategy improves network performance as compared to the
static data rate adaptation strategy in terms of packet delivery ratio
upto 15% and reduces packet loss due to interface queue overflow upto
14%, respectively.

1 Introduction

Due to mobility, lack of continues end-to-end connectivity, and dynamic network
topology, reliable data delivery becomes a challenging task in Mobile Ad Hoc
Networks (MANETs). When source node transmits data packets to the destination, any intermediate node can suffer from congestion due to limited resources.
Congestion will prompt high packet loss, long delays, and wastage of network
resources [1,2]. The reasons for packet loss may be due to node mobility, non-
availability of next hop nodes, interface queue overflow, and so on. As multiple
sources are sending frequent data, queue of forwarding node may overflow causing
packet drops which leads to degradation of the network performance. Therefore,
an efficient congestion control mechanism is of vital significance in networks like
MANETs. Existing conventional congestion control mechanisms are unable to

© ICST Institute for Computer Sciences, Social Informatics and Telecommunications Engineering 2017
J. Ferreira and M. Alam (Eds.): Future 5V 2016, LNICST 185, pp. 57–66, 2017.
DOI: 10.1007/978-3-319-51207-5_6

cope with the congestion in MANETs due to many inherent challenges like high node mobility and continuous changes in the topology of network.

Various techniques have been proposed to avoid packet loss, such as adapting alternate route, routing Using bypass concept, on-demand multicast, and so on. These schemes find the alternate path in case of congestion at nodes in the current path. However, run-time calculation of alternate path is overhead in the aforementioned schemes. Moreover, congestion reports used in alternate route adaptation techniques may be delayed which can effect the network performance. Similarly, sending control packets for congestion notification is an itself overhead for a congested network.

To avoid the packet loss due to interface queue overflow, there is a strong need of mechanism which adapts the data rate at source nodes based on the run-time network conditions. In this paper, we propose Data Rate Adaptation Strategy (DRAS) to avoid packet loss due to interface queue overflow. The proposed DRAS avoids congestion before it actually happens.

The rest of the paper is organized as follows. Section 2 presents related work. Section 3 describes our proposed technique. Performance evaluation is discussed in Sect. 4 and finally conclusion and future work is presented in Sect. 5.

2   Related Work

Packet loss due to congestion or queue overflow is a severe problem in MANETs. Based on the methodology used to avoid congestion, these schemes can be classified as alternate path based [3–5], data rate adaptation based and pausing control messages. In this work, we present the data rate adaptation based techniques, as our work is more related with this type of category. In the following, we discuss these schemes in details.

In [6], authors proposed a technique to control congestion in wireless sensor networks. Their technique calculates rate of sending packets, compare it to sending rate of parent node and downstream the smaller one. By rate of sending data, they get mean rate of packet generation of all nodes. They can reduce this rate if queue of node is full to minimize the congestion which results in packet loss reduction and hence, increases network performance. In [7] data rate is increased if 10 consecutive successful transmissions are done and decreased after 2 consecutive transmission failures. Success and failure is evaluated on the basis of received ACK packets. This scheme is not taking into account the reason of transmission failure. In [1,10,11] data rate is adapted for end-to-end congestion control. In end-to-end congestion control mechanism, ACK packets, sent by destination node, are used to communicate congestion message to the source nodes. A novel Rate-Based Congestion Control (RBCC) and EXplicit rAte-based flow ConTrol (EXACT) techniques are proposed, in [1,10] respectively, based on end-to-end congestion control mechanism. Technique which assists routing nodes named Explicit.

Another way of providing feedback to source node is explicit congestion notification (ECN) in header of the packet. Technique proposed in [12,13] are using the congestion notifications to avoid packet loss. In these techniques, once the data has reached, the router calculates the load factor for all the links and congestion region is identified on the basis of load factor. Moreover if calculated load factor is comparatively higher, then congestion status is updated by overwriting ECN bits. Technique proposed in [13] uses two ECN bits for improving results. In this technique, data rate is changed by a static factor. Another technique in which successful packet transmission and buffer threshold both are considered is additive increase and multiplicative decrease (AIMD) scheme [14] is proposed. In this scheme, for every successful packet transmission, data rate is increased with increasing parameter and continue until buffer threshold is received from other side and data rate is decreased when packet transmission failed. Failure of data transmission is measured by not receiving ACK packets. Due to congested route, delivery of ACK packet might be delayed that cannot represent the actual status of route.

Queue based data rate adaptation is done in [15–18]. These techniques uses queue length as a parameter to judge congestion. Probability of accessing the communication channel is calculated by each node based on the number of unsuccessful transmissions in [18]. In addition, each node receives a hello message periodically from its neighbors containing the channel access probability, transmission rate and the estimated traffic load. Reinforcement learning is used by each node to analyze the channel access probability. Thus, previous actions are used to decide either it is necessary to update the transmission rate or not. Negative impact of updating the transmission rate unnecessarily is mitigated using this technique. Furthermore, this technique also take in account the load on each node calculated by its queue length and then this information is used to decide whether to increase, decrease or keep its transmission rate. In this paper, no specific mechanism is discussed about factor by which data rate is increased or decreased. We assume it a static factor.

To adapt data rate at sender node, a technique is proposed in [15]. This technique is based on queue length analysis at intermediate nodes. One of the basis shortcomings of the aforementioned technique is that data rate is adaptation is fixed and static without analyzing the run-time network conditions.

To summarize, the proposed techniques are based on static data rate adaptation mechanism. Congestion notification messages are used to propagate information. Using congestion notification on congested route is an overhead for the network. To overcome the limitation of proposed techniques, we have proposed congestion avoidance strategy based on the dynamic rate adaptation for MANETs. Proposed strategy avoids the packet loss caused specifically due to interface queue (IFQ) overflow by dynamically adapting data rate at intermediate nodes on the basis of current queue size.

3   Data Rate Adaptation Strategy (DRAS)

Techniques using data rate adaptation at source node with static factor results in wastage of network resources. To avoid the negative influence of these techniques on network, a dynamic and adaptive data rate adaptation strategy is proposed to avoid packet loss caused due to interface queue overflow in MANETs. In this scheme, unlike feedback based rate adaptation, data rate is adapted dynamically by considering the network parameters having significant influence on data rate. In this paper, we have first identified the network parameters that are critical in order to adapt the data rate, and analyze their relationship to the network dynamics. Then, we discuss how such topology parameters affect the data rate.

3.1 Data Rate Adaptation and Network Parameters

In proposed technique we have identified various network parameters that have significant effect on data rate of nodes. Packet loss caused due to interface queue overflow is influenced by the factor of mobility, queue length and number of source nodes. We are using term queue length for consumed buffer space of a node.

Queue Length: Increased queue length results in high packet loss ratio so data rate is to be reduced with the increased queue length to avoid packet loss caused due to interface queue overflow. So, the relation between queue length and data rate is defined as:

$$DR \propto \frac{1}{QL},\tag{1}$$

where QL represents queue length and DR is data rate.

Queue length is being used as data rate adaptation parameter and varies between its minimum and maximum values. When queue is empty it has minimum value and maximum value denotes to value of queue when queue is full. On the basis of this maximum and minimum value, following formula helps us to find optimal rate adaption value at any node N with respect to queue length.

$$\rho = 1 - \frac{QL_N}{QL_{max}},\tag{2}$$

where QL_N is current queue length of node and QL_{max} is maximum queue size of a node.

According to above equation, the maximum value of queue length results lowest rate adaptation factor, i.e. 0, while the minimum value of queue length results in the higher rate adaptation factor with respect to queue length parameter, i.e. 1, where, DR is data rate and SN is number of source nodes.

Number of Sources: Number of source nodes is another parameter for data rate adaptation. When there are no source nodes, it has minimum value and a node has maximum number of source nodes if all one-hop neighbors are sending data to it. The relation between number of source nodes and data rate adaptation factor can be written as:

$$DR \propto \frac{1}{SN},\tag{3}$$

To find the optimal rate adaptation value at node N for avoiding packet loss caused due to queue overflow with respect to number of source nodes, we use the following equation:

$$\sigma = 1 - \frac{Sources_N}{nb_{max}}, \qquad (4)$$

where $Sources_N$ is current number of source nodes of a node N and nb_{max} is maximum possible source nodes of a node N.

According to above equation, the maximum value of number of source nodes results in lowest rate (i.e., 0), while the minimum value of number of source nodes results in the highest rate adaptation (i.e., 1).

Rate of Link Changes: Number of link changes shows the mobility of network. High mobility means node has less interaction time to exchange their packets. So, data rate should be higher that allows node to transfer its messages to other nodes in less time. This may over burden the queue of receiving nodes that may results in drop of packets. To avoid this problem, data rate is adopted keeping queue size of forwarding node in account. Mathematical relation between link changes and data rate can be written as:

$$DR \propto LC, \qquad (5)$$

where DR is data rate and LC is number of link changes.

Number of link changes is an important parameter for data rate adaptation. LC has minimum value when no node movement. Similarly, a node has maximum number of link changes if all 1-hop and 2-hop neighbors of that node have changed their position. The optimal value for data rate adaptation factor with respect to the link change rate is formulated as follows:

$$\varepsilon = \frac{LC_N}{N_O}, \qquad (6)$$

where LC_N is current number of link changes of a node and N_O is maximum possible number of link changes in a particular node's neighborhood.

According to above equation, the maximum value of number of link changes results in highest rate adaptation, i.e. 1, while the minimum value of number of link changes results in the lowest, i.e., 0.

3.2   Mathematical Model

We can combine the equations introduced so far into a mathematical model to compute data rate adaptation factor ϖ to avoid packet loss caused due to interface queue overflow. By combining Eqs. 2, 4, and 6, we obtain:

$$\varpi = \frac{(\alpha\rho + \beta\sigma + \gamma\varepsilon)}{(\alpha + \beta + \gamma)}, where \ \alpha + \beta + \gamma = 3. \qquad (7)$$

In Eq. 7, α, β, and γ are the weights assigned to each data rate adaptation parameter, discussed previously. As we are considering every parameter equally important for data rate adaptation, weights assigned to each parameter is equal, i.e. 1.

4   Performance Evaluation

In this section, we present the simulation setup, performance metrics, and simulation results to evaluate the proposed scheme in comparison to the protocol without rate adaptation mechanism. Proposed technique is evaluated on Network Simulator (NS2). The Optimized Link State Routing (OLSR) protocol is used as routing protocol. Table 1 shows the simulation parameters.

Table 1. Simulation parameters

Parameter	Value
Simulation time	1000 s
Number of nodes	50
Network size	1000 m × 1000 m
Transmission range	250 m
Packet size	512 b
Queue length	50
Mobility model	Random way point
Traffic type	Constant bit rate (CBR)
Max speed	1–10 m/s
Source-destination pairs	10–50%
Data rate	2–10 packets/sec

We have evaluated our proposed scheme referred as "DRA" in graphs in comparison to a protocol without any data rate adaptation strategy referred as "WRA". The proposed scheme is evaluated under two network performance parameters, i.e., under varying data rate and node speed.

4.1   Packet Delivery Ratio (PDR)

Figure 1a shows the effect of increasing data rate on PDR. As data rate increases, chances for packet loss are higher because queues of forwarding node overflows frequently. So, delivery probability decreases with increased data rate. The WRA scheme is not adapting any strategy to avoid packet loss, that is why it holds lower PDR as compared to the DRA scheme. As shown in figure, the DRA scheme has higher PDR comparatively as in this scheme data rate is adapted efficiently to avoid packet loss.

Similarly, Fig. 1b shows the effect of node mobility on PDR. Increased mobility results in frequent forwarding node queues overflow as nodes change their position frequently and forwarding nodes need to store packets for longer in their queues. So, the chances of packet loss become higher. As there is no data rate adaptation technique in the WRA scheme, with increasing mobility there is

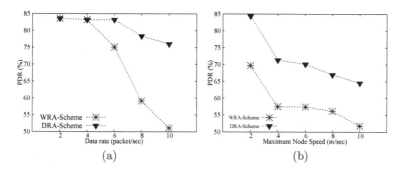

Fig. 1. Effect of data rate and node speed on PDR

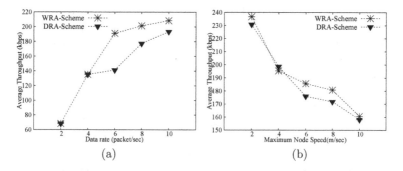

Fig. 2. Effect of data rate and node speed on throughput

more packet loss which cause lower PDR. As shown in Fig. 1b, the DRA scheme has more PDR as compared to WRA as in this scheme data rate is adapted efficiently to avoid packet loss.

4.2   Throughput

As data rate increases, throughput is also increasing because increased data rate means number of packets generated per unit time increases. As shown in Fig. 2a, WRA shows highest throughput as compared to the DRA scheme because it sends data with constant data rate without taking congestion in account. If congestion occurs, WRA do not have any mechanism to deal with such situation. On the other hand, proposed mechanism adds more parameters and adapts data rate in a better way. In adaptation phase, DRA is not utilizing whole bandwidth of the channel which results in decreased throughput that is why throughput in DRA scheme is slightly lower as compared to the WRA scheme.

Similarly, for increasing node mobility means nodes are mobile and change their position frequently. Forwarding nodes do not find the destination node and unable to deliver packets frequently that is why throughput decreases with increasing node mobility as shown in Fig. 2b. WRA shows highest throughput as compared to the DRA scheme because it sends data with constant data rate without taking congestion into the account. Due to data rate adaptation mechanism in the DRA scheme, the data rate is decreased at the source node to avoid packet loss causing under utilization of the channel bandwidth, hence lower throughput in this case.

4.3 End-to-End Delay

End-to-end delay increases with increased number of packet loss as packet loss results in large number of re-transmissions, hence increases end-to-end delay. In WRA scheme, as packet loss is high, so they result in long end-to-end delays as shown in Fig. 3a. Proposed scheme reduces delivery delay because data rate is efficiently adapted which low packet loss. Moreover, as DRA scheme is adapting data rate to avoid packet loss which means smaller number of packets in network will be communicated, hence avoid congestion. As congestion is avoided, so end-to-end delay also decreases in proposed scheme due to smaller number of re-transmissions for lost packets.

Similalry, Fig. 3b shows the impact of mobility of average end-to-end delay. Due to high mobility, data is not delivered to distant nodes. Where as, data is delivered speedily to nearer nodes that is why end-to-end delay decreases with increased mobility. Proposed scheme reduces delivery delay as compared to the WRA scheme as shown in Fig. 3b because data rate is efficiently adapted which reduces packet loss. Hence, large number of re transmissions are not required which effects end-to-end delay positively.

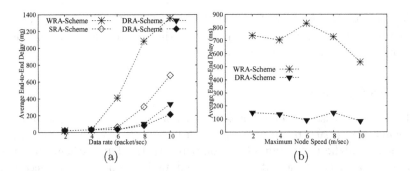

Fig. 3. Effect of data rate and node speed on end-to-end delay

4.4 Packet Loss

As shown in Fig. 4a packet loss due to interface queue overflow increases with the increased data rate because queues of forwarding node overflow due to high data rate. Moreover, Fig. 4a shows that packet loss in WRA scheme are higher as this scheme does not consider any strategy to avoid packet loss. But on the other hand, DRA scheme adapts data rate efficiently by considering mobility

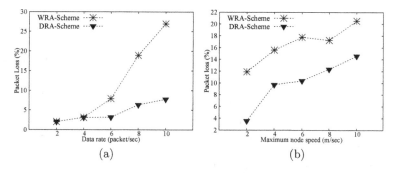

Fig. 4. Effect of data rate and node speed on packet loss

factor and number of source nodes along with queue length which causes smaller packet loss in proposed scheme.

To analyze the impact of node mobility on packet loss, Fig. 4b packet loss with increasing node mobility. For increasing node mobility, forwarding node queues are overflow as nodes change their position frequently and forwarding nodes need to store packets for longer in their queues. Moreover, Fig. 4b shows that packet loss in WRA scheme are higher as this scheme does not consider any strategy to avoid packet loss. But on the other hand, the DRA scheme adapts data rate efficiently based on multiple network factors to avoid packet loss.

5   Conclusion and Future Work

In this paper, we proposed a technique to adapt data rate of the sender node based on the run time network conditions around forwarding nodes using different network dynamics. Proposed technique is based on the analysis of queue length of the forwarding nodes. We have simulated our proposed technique in NS-2 and achieved better results in terms of packet delivery ratio and average end-to-end delay in comparison to the static rate adaptation technique.

As currently, the data rate is adapted at source nodes by shifting the congestion effect immediately from forwarding nodes to the source node. In future, we plan to adapt the data rate at intermediate nodes based on the mentioned parameters by buffering the data packets in queues up to bearable threshold, and then gradually shift the effect of rate adaptation to the source nodes. We also plan to evaluate the proposed scheme in VANETs scenario to under different performance metrics to check the compatibility and scalability of the proposed scheme in high dynamic environment.

Acknowledgment. The work reported in this paper has been partially supported by Higher Education Commission (HEC), Pakistan, and the University of Malaya under grant RG325-15AFR.

References

1. Chen, K., Nahrstedt, K., Vaidya, N.: The utility of explicit rate-based flow control in mobile ad hoc networks. IEEE Wireless Communications and Networking Conference, vol. 3, pp. 1921–1926 (2004)
2. Deshmukh, V.S., et al.: A survey of congestion control in proactive source routing protocol in mobile ad hoc networks. COMPUSOFT **3**(12), 1350 (2014)
3. Yaghmaee, M.H., Adjeroh, D.A.: Priority-based rate control for service differentiation and congestion control in wireless multimedia sensor networks. Comput. Netw. **53**(11), 1798–1811 (2009)
4. Hussain, A., Zia, K., Farooq, U., Ahmed, S.: PLP-A packet loss prevention technique for partitioned MANETs using Location-Based Multicast (LBM) Algorithms. In: National Conference on Emerging Technologies (2004)
5. Kumaran, T.S., Sankaranarayanan, V.: Early congestion detection and adaptive routing in MANET. Egypt. Inform. J. **12**(3), 165–175 (2011)
6. Ee, C.T., Bajcsy, R.: Congestion control and fairness for many-to-one routing in sensor networks. In: Proceedings of the 2nd ACM International Conference on Embedded Networked Sensor Systems, pp. 148–161 (2004)
7. Camp, J., Knightly, E.: Modulation rate adaptation in urban and vehicular environments: crosslayer implementation and experimental evaluation. IEEE/ACM Trans. Netw. **18**(6), 1949–1962 (2010)
8. Chevillat, P., Jelitto, J., Barreto, A.N., Truong, H.L.: A dynamic link adaptation algorithm for IEEE 802.11 a wireless LANs. In: IEEE International Conference on Communications (ICC 2003), vol. 2, pp. 1141–1145 (2003)
9. Xi, Y., Kim, B.S., Wei, J.B., Huang, Q.Y.: Adaptive multirate auto rate fallback protocol for IEEE 802.11 WLANs. In: IEEE Military Communications Conference (MILCOM 2006), pp. 1–7 (2006)
10. Senthilkumaran, T., Sankaranarayanan, V.: Dynamic congestion detection and control routing in ad hoc networks. J. King Saud Univ. Comput. Inf. Sci. **25**(1), 25–34 (2013)
11. Zhai, H., Chen, X., Fang, Y.: Rate-based transport control for mobile ad hoc networks. In: IEEE Wireless Communications and Networking Conference, vol. 4, pp. 2264–2269 (2005)
12. Manikandan, K., Durai, M.A.S.: Active queue management based congestion control protocol for wireless networks. Int. J. Enterp. Netw. Manag. **6**(1), 30–41 (2014)
13. Tiwari, S.K., Rana, Y.K., Jain, A.: An ECN approach to congestion control mechanisms in MANETs. Netw. Complex Syst. **4**(6), 18–22 (2014)
14. Mishra, T.K., Tripathi, S.: Explicit throughput and buffer notification based congestion control: a cross layer approach. In: Eighth International Conference on Contemporary Computing (IC3), pp. 493–497 (2015)
15. Thakur, S., Gupta, M.: Mitigating congestion using data rate control for MANETs (2014)
16. Rahman, K.C., Hasan, S.F.: Explicit rate-based congestion control for multimedia streaming over mobile ad hoc networks. Int. J. Electr. Comput. Sci. IJECS-IJENS **10**(04), 28–40 (2010)
17. Benslimane, A., Rachedi, A.: Rate adaptation scheme for IEEE 802.11-based MANETs. J. Netw. Comput. Appl. **39**, 126–139 (2014)
18. Al-Saadi, A., Setchi, R., Hicks, Y., Allen, S.M.: Multi-rate medium access protocol based on reinforcement learning. In: IEEE International Conference on Systems, Man and Cybernetics (SMC), pp. 2875–2880 (2014)

Security and Applications
of Vehicular Communication

Low-Cost Vehicle Driver Assistance System for Fatigue and Distraction Detection

Sandra Sendra[1,2], Laura Garcia[2], Jose M. Jimenez[2],
and Jaime Lloret[2(✉)]

[1] Signal Theory, Telematics and Communications Department (TSTC),
Universidad de Granada, C/ Periodista Daniel Saucedo Aranda,
s/n, 18071 Granada, Spain
ssendra@ugr.es
[2] Integrated Management Coastal Research Institute,
Universidad Politecnica de Valencia, C/ Paraninf, 1, 46730 Grao de Gandia, Spain
laugarg2@teleco.upv.es, {jojiher,jlloret}@dcom.upv.es

Abstract. In recent years, the automotive industry is equipping vehicles with sophisticated, and often, expensive systems for driving assistance. However, this vehicular technology is more focused on facilitating the driving and not in monitoring the driver. This paper presents a low-cost vehicle driver assistance system for monitoring the drivers activity that intends to prevent an accident. The system consists of 4 sensors that monitor physical parameters and driver position. From these values, the system generates a series of acoustic signals to alert the vehicle driver and avoiding an accident. Finally the system is tested to verify its proper operation.

Keywords: Low-cost sensors · Vehicular technology · Driver assistance system · Fatigue episodes · Distraction detection · Sensing system

1 Introduction

As years go by, vehicular technology has been improving to satisfy the needs of its users. Most of this technology is based on acquiring data to improve the performance of the car and enhance its safety. To do this, cameras, sensors and Global Positioning System (GPS) technology are used. These devices are employed on active and passive safety systems [1] such as Antilock Brake System (ABS), Electronic Stability Control (ESP), power steering, airbags or seatbelts.

As it is shown in the annual report on road casualties provided by the Spanish Government [2], the number of victims in traffic accidents has decreased since 1989. The decrease of victims has happened even when the number of cars in the country has increased. One of the main reasons of the decrease in mortality is the usage of different kind of sensors that increase the safety of the car. Many of these sensors are created to alert the vehicle driver of an imminent collision [3].

Despite the decrease of traffic fatalities, the number of deaths in traffic accidents is quite disturbing. In 2014, the number of deaths caused by traffic accidents in Spain was

© ICST Institute for Computer Sciences, Social Informatics and Telecommunications Engineering 2017
J. Ferreira and M. Alam (Eds.): Future 5V 2016, LNICST 185, pp. 69–78, 2017.
DOI: 10.1007/978-3-319-51207-5_7

1.688 people [4]. Many deaths are caused by the intake of toxic substances such as alcohol or drugs. Even prescription drugs may generate side effects that affect driving. Fatigue can be caused by the aforementioned substances and sleep deprivation, hot weather or driving long distances, among others.

Fatigue presents a wide range of symptoms [5]. The driver's vision turns blurry and the blinking rate gets higher. It also affects the behavior causing anxiety and making the driver more irritable. Fatigue increases the number of movements the driver does to accommodate as well as other type of movements such as tapping the wheel. Finally, Fatigue increases the time the driver takes to react in a dangerous situation. For this reason, it is important to have systems able to detect the aforementioned symptoms. This can help to increase the driver's safety, but most of the currently developed solutions are focused on eye movement and face detection [6].

This paper presents a low-cost vehicle driver assistance system to detect episodes of fatigue and distraction in drivers.

The rest of the paper is organized as follows. Section 2 presents the related work. Section 3 explains the system operation and presents the algorithm used to detect and classify the alarm level. This section also shows the system tests. Finally, Sect. 4 presents the conclusion and future work.

2 Related Work

We can find lots of previous works about how to monitor and study fatigue in drivers and sleepiness while driving. This section shows some of these works.

In general, these works can be divided into two groups. On the one hand, we find proposals based on facial recognition. Sigari et al. [7] present an interesting review of driver face monitoring systems for fatigue and distraction detection where the general structure of these systems is discussed.

Kutila et al. [8] describes a facility for monitoring the distraction of a driver and presents some early evaluation results. They present a module that is able to detect the driver's visual and cognitive workload by fusing stereo vision and lane tracking data, running both rule–based and support-vector machine (SVM) classification methods. The module has been tested with data from a truck and a passenger car. The results show over 80% success in detecting visual distraction and a 68–86% success in detecting cognitive distraction, which are satisfactory results.

Rezaei and Klette [9] present a research that develops optimum values of Haar-training parameters to create a nested cascade of classifiers for real-time eye status detection. They present the unique features of their robust training database that significantly influenced the detection performance. Their systems have been implemented and tested in real-world with satisfactory results.

Mbouna et al. [10] present visual analysis of eye state and head pose (HP) for continuous monitoring of alertness of a vehicle driver. The proposed scheme uses visual features such as eye index (EI), pupil activity (PA), and HP to extract critical information on no alertness of a vehicle driver. Using a support vector machine (SVM) classifies a sequence of video segments into alert or nonalert driving events.

Their experimental results show that their proposed scheme offers high classification accuracy with acceptably low errors and false alarms for people of various ethnicity and gender in real road driving conditions.

Wahlstrom et al. [11] present a project that involves the use of a dashboard mounted camera to monitor the direction a driver is looking. They accomplish the project by using the Framework for Processing Video (FPV), developed at the University of Minnesota by Osama Masoud. Their software uses the relative positions of the eyes and pupils to make statements about the gaze direction.

Cherrat et al. [12] propose a multifunction system based on intelligent sensors and cameras. The system is mainly based on learning systems for face recognition based on advanced algorithms Viola and Jones, PCA and management of drivers profiles based on preferences to provide the following features: early detection of sleep, unconsciousness and poor driver behavior, security against theft of vehicles, driver comfort and control and sharing of traffic information in real time between the conductors.

On the other hand, there are some other works that use other techniques to detect fatigue episodes in drivers. In this sense, we can find works as the one, presented by Dong et al. [13] who review the state-of-the-art technologies for driver inattention monitoring such as distraction and fatigue. In their work, authors summarize these approaches by dividing them into the following five different types of measures: (1) subjective report measures; (2) driver biological measures; (3) driver physical measures; (4) driving performance measures; and (5) hybrid measures. Authors think that the hybrid measures are believed to give more reliable solutions compared with single driver physical measures or driving performance measures, because the hybrid measures minimize the number of false alarms and maintain a high recognition rate, which promote the acceptance of the system.

All of these solutions require powerful systems to process all the generated data. Our proposal tries to implement a very cheap solution that can be used in any kind of car. It is based on the measurement of physical parameters and its main task is generating a fast acoustic alarm to prevent the driver and avoid possible accidents.

3 System Description

When implementing a sensing system able to be used by everybody, we have to provide low-cost solutions. This section presents the design of our proposal. It is based on several sensors which are in charge of collecting data from the vehicle driver and generates alert signals to avoid the driver sleeps and may suffer an accident.

3.1 Overall Explanation

Our low-cost vehicle driver assistance system is based on economic devices and it has been designed to be used in any kind of car. A simple processing unit is in charge of collecting data from several sensors. From the gathered data, the system will generate a kind of alarm in order to prevent the driver.

As Fig. 1 shows, the car is endowed of various sensors. In this case, we use 2 pressure sensors and 2 temperature sensors installed in the steering wheel (duplicate system for monitoring both hands). It also includes a proximity sensor based on a LDR installed in the headrest which will detect if the driver separates the head a considerable distance to the headrest. If the received light is too big, this could imply that the driver is starting to get drowsy and can nodding off.

Our Low-cost vehicle driver assistance system contains a buzzer near to the ear driver that will generate an acoustic alarm in case of detecting a fatigue state and driver distraction.

Finally, we have installed 2 more elements. The first one is a shock sensor that detects collisions or sudden braking and a push-button placed on the steering wheel, to interact with the driver.

The sensors are wired to the electronic board. The system performs the sensors' sampling every second and stores the results on a micro-SD card for possible post-processing task, in case of accidents or future studies. The electronic board checks that the values collected by the sensors remain within the thresholds considered as normal. When any of these sensors exceeds any of these thresholds, the proposed algorithm (see Fig. 2) try to define whether it is a false alarm or sensors have registered a possible situation of the vehicle driver fatigue or distraction.

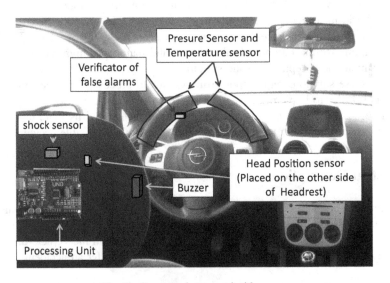

Fig. 1. Sensors placement inside a car.

As algorithm of Fig. 2 shows, the low-cost vehicle driver assistance system is continuously monitoring the sensors' activity. When it detects that some sensor has exceeded the specified thresholds, the system will try to define how many sensors are recording an anomalous behavior. As a function of this value, the system will generate a sound of a specific frequency and a pattern. For example, if the system

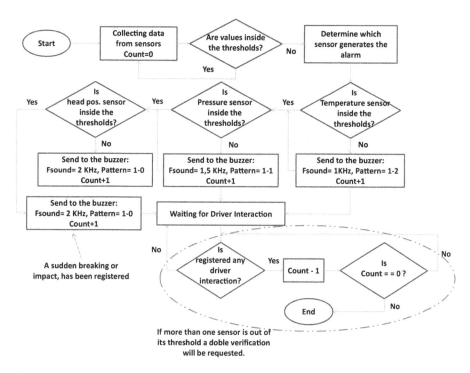

Fig. 2. Algorithm used by our low-cost vehicle driver assistance system for detecting fatigue and distraction episodes and generate the alarms.

detects that the temperature is the only parameter that has exceeded the threshold, the buzzer will emit a sound of 1 kHz each 2 s. If the system detects that the temperature and pressure sensor have exceeded their thresholds, the buzzer will emit a sound of 1.5 kHz each seconds. Finally, if all sensors have exceeded their thresholds, the buzzer will emit a continuous sound of 2 kHz. If a sudden breaking or impact has been registered the buzzer will also emit a continuous sound of 2 kHz.

To disable the alarm status and return the system to the sensing phase, the system requires the driver interaction by pushing the button placed on the steering wheel. If more than one sensor is out of its threshold a double verification will be requested.

3.2 Processing Unit

The system is based on a small electronic board based on the ATmega328 microprocessor. It has 14-pin digital inputs/outputs (6 of them can be used as PWM outputs), 6 analog inputs, a crystal oscillator of 16 MHz. This device can be powered from a computer via the USB connection or using batteries. Figure 3 shows the processing unit. Unlike some previous models, this version allows to program the device without any FTDI USB-to-serial converter. In this case, the Board ONE uses a programmed Atmega16U2 chip which is used as a USB to serial converter.

Fig. 3. Processing unit used in the design of our low-cost vehicle driver assistance system

3.3 Sensors Used by the Low-Cost Vehicle Driver Assistance System

The system is basically composed by 4 sensors and 2 actuators. This subsection presents the main features of each one.

Sensor for Measuring the Pressure of Hands on the Wheel: A Force Sensitive Resistor (FSR) [14] is a sensor that allows detecting and measuring physical pressure, squeezing and weight. This sensor is based on a resistor that changes its resistive value (Ω) depending on the pressure it registers. However, FSRs are not indicated to accurate measurements because its accuracy can vary up to 10% from sensor to sensor. So, in our case, the FSR is used to detect a variation that exceeds a threshold. Figure 4a shows the FSR sensor. The FSR 402 is able to support values of pressure from 0 to 100 Newtons which correspond to a resistance range from Infinite/open circuit (absence of pressure) to100 KΩ (light pressure). Since its behavior is like a resistor, it can be used as a part of a resistive voltage divider. In this way, the processing unit will register this pressure variation as a voltage value. Figure 4b shows the output voltage of this sensor when it is combined with a resistance of 1 kOhm and a power supply of 3.3 V.

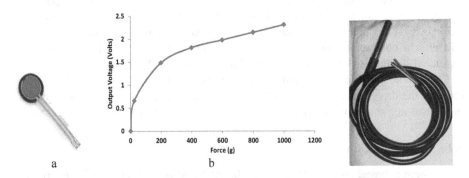

a   b

Fig. 4. Force Sensitive Resistor and its behavior.   **Fig. 5.** PT100

Temperature Sensor: To develop the temperature sensor, it is used a resistance temperature detector (RTD). It is a resistance based on the variation of the resistance of a conductor with the temperature sensor (See Fig. 5). Their behavior is very simple. When heating a metal, it is generated a higher thermal agitation that provokes the dispersion of more electrons reducing its average speed. This causes that resistance increase. The behavior of this sensor can be modeled by Eq. 1, where R is the value of resistance in (Ω), R_0 is the resistance at reference temperature T_0, Δt is the temperature deviation with respect to T_0 and α is the is the conductor temperature coefficient at 0 °C. This sensor presents a linear behavior in wide temperature ranges.

$$R = R_0(1 + \alpha \cdot \Delta t). \tag{1}$$

Head Position Sensor: This sensor is based on a light-dependent resistor (LDR). In this case, its resistance decreases when increasing incident light intensity. We can use this sensor combined with other resistance forming a voltage divider. If we use the LDR placed at the bottom of the voltage divider, it will give us the maximum voltage when the LDR is in total darkness, because it is having the maximum resistance to the current flow. In this situation registers the maximum value. However, if the LDR is placed at the top of the voltage divider, the result is the opposite.

Figure 6 shows the head position sensor and its position on the headrest. The particularity of this small board is that it allows controlling the sensitivity of this sensor.

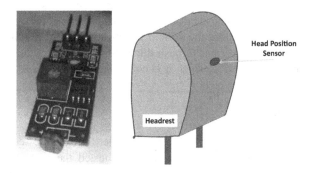

Fig. 6. Head position sensor and its position in the headrest

Shock Sensor: It is based on the Gaoxin SW-18010P vibration switch and allows using our system to detect impacts, shocks or shaking. When the sensor detects a jolt, the sensor generates a low level output signal. Figure 7 shows the module used in this design. The sensor consists of a terminal that forms a center post and a second terminal that is a spring that surrounds the center post. If a sudden breaking or impact with sufficient force is transferred to the sensor, the terminal consisting of the spring moves and shorts both terminals together. The connection between the terminals is short but the central unit is able to detect this breakdown voltage. The sensor position is also important. In our case, it is place in the forward direction of the vehicle (Fig. 1).

Fig. 7. Shock sensor   **Fig. 8.** Buzzer module placed near to the driver's ear.   **Fig. 9.** Push-button for alarm verification

In order to alert the driver, our low-cost vehicle driver assistance system contains a buzzer. It is a KY-006 Small passive buzzer module (See Fig. 8). It is composed by a piezo-resistive element and a resonance box that permit generating audio signals from 1.5 kHz to 2.5 kHz. As Fig. 1 showed, this element is placed near to the driver's ear. It is used as an alarm element which is in charge of generating an acoustic alarm to alert the driver if a fatigue status of distraction is detected. The output frequency can be configured by programing.

The last important element is a verification system. It is a simple push-button that requires the driver interaction (Fig. 9). After detecting a dangerous situation, our driver assistant system will emit an acoustic alarm to alert the driver and prevent he/she sleeps while driving. In order to restore the system which implies that the driver is paying attention to the road, the driver has to push this button. After that, the system will go back to the sensing phase during which it is collecting data.

3.4 System Simulation

Finally, it is interesting to check if the system is able to detect the sensors value and generate the correct alarms. To check it, we have tested the system during 4 min. During this time, we have forced the sensors to change their values (See Fig. 10) and to

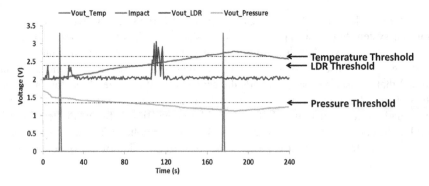

Fig. 10. Sensors monitoring

exceed the correspondent threshold. As we can see in Fig. 11, the low-cost vehicle driver assistance system is registering the alarm level and as the user pushes the button to disable the alarm, the alarm level decreases.

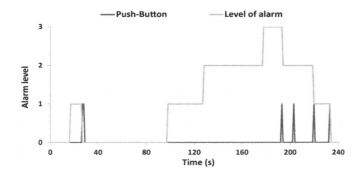

Fig. 11. Alarm generation as a function of sensors value

4   Conclusion and Future Work

When implementing a sensing systems and we want to offer these systems to as many people as possible, it is needed to think with low-cost system that allows developing accurate designs. This paper presents a low-cost vehicle driver assistance system for detecting fatigue episodes and distraction situations of a vehicle driver.

The system is based on an electronic board that acts as processing unit and 4 sensors (person temperature, pressure on the steering wheel, a light sensor placed on the headrest and a shock sensor that detect sudden breakings and impacts). The low-cost vehicle driver assistance system is able of classifying the alarms as a function of the number of sensors that have registered values out of the thresholds. It also contains 2 actuators, i.e., a buzzer and a push-button for interacting with the driver. After generating an alarm the driver has to disable it pushing the button.

This is the initial version of prototype. In future works, we want to include a small Global Positioning System (GPS) receiver to gather the vehicle position and a wireless technology module to interact with the smartphones in order to create a more autonomous system able to generate an emergency call via the smartphone. Finally, we would like to integrate the system in a more complex and energy efficient network [15, 16] with the rest of sensors embedded in a car to generate more accurate responses.

Acknowledgments. This work has been partially supported by the "Programa para la Formación de Personal Investigador – (FPI-2015-S2-884)" by the "Universitat Politècnica de València".

References

1. Mapfre Foundation. (Online Article) Seguridad activa y pasiva. www.seguridadvialen laempresa.com/seguridad-empresas/actualidad/noticias/seguridad-vial-activa-y-pasiva-2.jsp. Accessed 25 Aug 2016
2. Dirección general de tráfico, Ministerio del Interior, Spanish Government. (Online Article) Las principales cifras de la siniestralidad vial. España 2014, p. 21 (2014). http://www.dgt.es/es/seguridad-vial/estadisticas-e-indicadores/publicaciones/. Accessed 25 Aug 2016
3. Fukuhara, H.: Vehicle collision alert system. US Patent 5355118 A, 11 Oct 1994
4. Dirección general de tráfico, Ministerio del Interior, Spanish Government. (Online Article) Anuario estadístico de accidentes 2014, p. 10 (2014). http://www.dgt.es/es/seguridad-vial/estadisticas-e-indicadores/publicaciones/anuario-estadistico-general/. Accessed 25 Aug 2016
5. Dirección general de tráfico, Ministerio del Interior, Spanish Government. (Online Article) Otros factores de riesgo: La fatiga. http://www.dgt.es/PEVI/documentos/catalogo_recursos/didacticos/did_adultas/fatiga.pdf. Accessed 25 Aug 2016
6. Seeing machines web page. https://www.seeingmachines.com/. Accessed 25 Aug 2016
7. Sigari, M.H., Pourshahabi, M.R., Soryani, M., Fathy, M.: A review on driver face monitoring systems for fatigue and distraction detection. Int. J. Adv. Sci. Technol. 64, 73–100 (2014). http://dx.doi.org/10.14257/ijast.2014.64.07
8. Kutila, M., Jokela, M., Markkula, G., Romera Rue, M.: Driver distraction detection with a camera vision system. In: 14th IEEE International Conference on Image Processing (ICIP 2007), San Antonio, TX, USA, 16–19 September 2007. doi:10.1109/ICIP.2007.4379556
9. Rezaei, M., Klette, R.: 3D cascade of classifiers for open and closed eye detection in driver distraction monitoring. In: Real, P., Diaz-Pernil, D., Molina-Abril, H., Berciano, A., Kropatsch, W. (eds.) CAIP 2011. LNCS, vol. 6855, pp. 171–179. Springer, Heidelberg (2011). doi:10.1007/978-3-642-23678-5_19
10. Mbouna, R.O., Kong, S.G., Chun, M.G.: Visual analysis of eye state and head pose for driver alertness monitoring. IEEE Trans. Intell. Transp. Syst. 14(3), 1462–1469 (2013). doi:10.1109/TITS.2013.2262098
11. Wahlstrom, E., Masoud, O., Papanikolopoulos, N.: Vision-based methods for driver monitoring. In: Proceedings of the International Conference on Intelligent Transportation Systems, vol. 2, pp. 903–908 (2003)
12. Cherrat, L., Ezziyyani, M., El Mouden, A., Hassar, M.: Security and surveillance system for drivers based on user profile and learning systems for face recognition. Netw. Protoc. Algorithms 7(1), 98–118 (2015). doi:http://dx.doi.org/10.5296/npa.v7i1.7151
13. Dong, Y., Hu, Z., Uchimura, K., Murayama, N.: Driver inattention monitoring system for intelligent vehicles: a review. IEEE Trans. Intell. Transp. Syst. 12(2), 596–614 (2011). doi:10.1109/TITS.2010.2092770
14. Force Sensitive Resistor features. http://www.trossenrobotics.com/productdocs/2010-10-26-DataSheet-FSR402-Layout2.pdf. Accessed 25 Aug 2016
15. Louiza, M., Samira, M.: A new framework for request-driven data harvesting in vehicular sensor networks. Netw. Protoc. Algorithms 5(4), 1–18 (2013)
16. Yao, H., Si, P., Yang, R., Zhang, Y.: Dynamic spectrum management with movement prediction in vehicular ad hoc networks. Ad Hoc Sens. Wirel. Netw. 32(1), 79–97 (2016)

Implementation of Security Services for Vehicular Communications

Daniel Duarte[1,2], Luis Silva[1,2(✉)], Bruno Fernandes[1], Muhammad Alam[1], and Joaquim Ferreira[1,3]

[1] Instituto de Telecomunicações, Aveiro, Portugal
brunofernandes@ua.pt, alam@av.it.pt
[2] DETI, Universidade de Aveiro, Aveiro, Portugal
daniel.duarte@ua.pt, luissilva@av.it.pt
[3] ESTGA, Universidade de Aveiro, Aveiro, Portugal
jjcf@ua.pt

Abstract. Over the last years, there has been a considerable development in the field of vehicular communications (VC) so as to satisfy the requirements of Intelligent Transportation Systems (ITS). Standards such as IEEE 802.11p and ETSI ITS-G5 enable the so called Vehicular Ad-Hoc Networks (VANETs). Vehicles can exploit VANETs to exchange information, such as alerts and awareness information, so as to improve road safety. However, due to the expected popularity of ITS, VANETs could be prone to attacks by malicious sources. To prevent this, security standards, such as IEEE 1609.2, are being developed for ITS. In this work, an implementation of the required cryptographic algorithms and protocols for the transmission of secure messages according to the IEEE 1609.2 standard is presented. The implemented security protocols are then integrated into an existing WAVE-based system and tested in a real scenario to evaluate the performance impact on safety-related communications, in particular, the overhead that is caused by the process to sign/verify a digital message.

Keywords: Vehicular Communications · Intelligent Transportation Systems · Security · IEEE 1609.2

1 Introduction

During the past decades, the volume and density of road traffic has increased significantly, specially in developing countries such as India and Brazil. According to [1], in 2010 the number of vehicles in the world has reached to 1.015 billion, with an approximate ratio of 1:7 cars per person. This significant growth lead to an increase in number of accidents and traffic injuries, with negative impacts on the economy and in the quality of people's lives [2]. In order to tackle this problem new systems, commonly known as Intelligent Transportation Systems (ITS), are being developed. In ITS, Dedicated Short Range Communications (DSRC), on a dedicated spectrum in the 5.9 GHz band, are employed to enable wireless

© ICST Institute for Computer Sciences, Social Informatics and Telecommunications Engineering 2017
J. Ferreira and M. Alam (Eds.): Future 5V 2016, LNICST 185, pp. 79–90, 2017.
DOI: 10.1007/978-3-319-51207-5_8

communication between vehicles and infrastructure. New applications can then be developed to exploit vehicle to vehicle (V2V) [8] and vehicle to infrastructure (V2I) [9] communications to improve road traffic's safety and efficiency. To this end, a group of new standards that specifies and standardizes several aspects of the aforementioned communications (e.g. physical and medium access control layers, data structures, security, etc.) has been defined. The most well known standards are the Wireless Access in Vehicular Environments (WAVE) in the USA and ETSI ITS-G5 in Europe.

Due to the high popularity and scale that ITS can attain, communications in vehicular networks, more specifically in Vehicular Ad-Hoc Networks (VANETs), are expected to be prone to security threats since these could be exploited by people with malicious intents (e.g. redirect traffic flow and spread false information etc.). Eavesdropping, message manipulation and replay attacks are examples of attacks that may target a VANET. Hence, an architecture, a set of interfaces and services that enable secure V2V and V2I wireless communications are also included in WAVE and ETSI ITS-G5 group of standards [3].

Security objectives and solutions are well defined for computer-based architectures in general but for vehicular environments the approach needs to be different due to its distinct properties and requirements [5]. For example, robustness and time constraints in VC are demanding. Vital functions for driving and/or alerts sent by other vehicles must be correctly processed in real-time; delays or errors might lead to vehicle malfunctions, bad driving decisions, or other occurrences that could cause physical damages and injuries. The small embedded computers found in vehicles may not have the necessary memory and performance for cryptographic operations without affecting the aforementioned functions. Moreover, since a car typically has a life-time of at least 10 years, an upgrade of the computational resources may not be possible and thus, it is important to assure all security requirements for cars' life time-frame. This work aims to assess the requirements of such security mechanisms for vehicular environments as well as to evaluate the performance impact that such mechanisms may induce in communications. To that end, a software implementation of the IEEE P1609.2 [5], the current security standard embedded in WAVE, will be presented and evaluated in a real world scenario.

This paper is structured as follows. Section 2 provides a brief overview of main features of the IEEE 1609.2 standard. Section 3 presents the software implementation of IEEE 1609.2 algorithms. Section 4 presents the integration of the implemented software algorithms into an existing WAVE-based framework. In Sect. 5 experimental scenarios to evaluate the performance of the implementation are described and the results presented. The paper concludes with the discussion of the obtained results in Sect. 6.

2   IEEE 1609.2 Brief Overview

Our implementation is based on the IEEE security standard 1609.2 D17 [6]. The required algorithms that the standard forces to be used in order to provide adequate security are the public key algorithms based on elliptic curves. The only symmetric algorithm that the standard refers is the AES-128 to be used with the ECIES public key algorithm. Therefore the standard leads to the following specification in terms of security algorithms:

- Digital Signatures using ECC over prime fields, ECDSA with NIST Fp curves;
- Encryption using ECC, ECIES;
- Hash Algorithms - SHA-1 and SHA-256;
- Symmetric scheme AES.

The ECDSA should support a 224-bit and a 256-bit key, the ECIES a 256-bit and the AES a 128-bit key implementation. The standard also refers the creation of specific certificates called WAVE-Certificates which are more compact for performance reasons. Although this is the most recent standard for security in VC it is very poor regarding its content. Most of the topics are still open, such as how certificates should be shared among elements on the road and which are the secure protocols to correctly use with the ECIES and AES algorithms. The main focus of this standard is to allow authenticity by using the ECDSA algorithm.

3   IEEE 1609.2 Implementation

3.1   System Architecture

In this section, a software implementation of the security services is proposed. C programming language was chosen due to its performance and available open-source libraries. OpenSSL [4] was picked as the auxiliary library to implement the security services and the cryptography engine since it is a free and open-source c/c++ library tailored for cryptographic operations and algorithms.

The implementation architecture was divided into 3 main modules: one to handle all the secure protocols, another responsible for the cryptographic algorithms and finally, one containing all the secure keys and manufacture values. The module that deals with security protocols handles requests and interprets the responses from the cryptography engine. It is also responsible for the requests to the locally stored keys and provides them to the cryptographic engine. This module is the interface between higher layers (e.g. applications) and the process of securing data. The private and public keys, along with certificates, are stored in the module that contains all the manufacture values and secure keys. In addition to these 3 modules, a Global Positioning System (GPS) module was used to support the application by providing location and time information. The proposed security model is based on the conceptual security model referred in [7] which is illustrated in Fig. 1. The engine which performs all the cryptographic algorithms was implemented with the OpenSSL library and it is capable of performing Elliptic Curve Digital Signature Algorithm (ECDSA) 224 and 256, ECIES, Hash algorithms, certificate generation and verification.

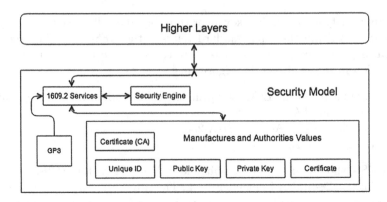

Fig. 1. Security model

Software Resources: The following set of software and libraries were used in the implementation.

Hash Table: A hash table is an abstraction of a common array in which contents can be accessed using keys instead of indexes. These tables allow unordered data to be searched and accessed very quickly.

Serialization: Serialization is the process of transforming data structures into a buffer of data, suitable for storage or transmission (the reverse process is also possible). This serialization mechanism is used to process structures and packets defined by the 1609.2 standard.

Sockets: When the connection between two programs or two machines is needed, the use of a mechanism called sockets can be used. Sockets will be used to interconnect certain software processes as well as the security module with the existing WAVE-based system.

GPSd: The GPSd is a daemon for GPS devices which interacts with various GPS brand devices through Universal Serial Bus (USB) serial interface enabling users to easily interact with different GPS devices and obtain GPS information. The GPSd is used in this implementation in order to get the time stamp and location to be inserted in the 1609.2 secure packet.

Cryptographic Material: The keys and certificates that should be inserted into the On Board Unit (OBU) are referred as crypto-material and are stored in manufacture time. We assumed to have a fully Public-Key Infrastructure (PKI) deployed, thus, algorithms to share certificates and manage certificate-revocations were not considered. Each car has a 224-bit and a 256-bit key-pair length and a certificate which can be chosen depending on the required algorithm. Thus, in this implementation, each vehicle contains the following crypto-material:

– A private and public key associated with the vehicle;
– A certificate containing the vehicle public key;
– The certificates that belong to all other cars.

Certificate hashes are included in the transmitted messages so as to avoid including the entire certificate. The SHA-256 is the chosen algorithm to perform this operation, but instead of adding the entire hash (256 bit) to the packet only the 8 less significant bytes are added [6]. Each car also contains a table of hashes, identifying all the stored certificates and its corresponding hash.

Data Flow: The way the signature generation/verification process works is defined by several services that handle multiple protocols. Different types of messages are going to be signed depending on the type of information that is going to be sent over the network. Each time a connection is requested to the security services, the flow of data is different depending on whether the request is to sign or to verify a message. The program starts by trying to establish a connection to the GPSd Daemon and after it is connected it creates a client/server connection with the facilities layer. Then the system awaits the reception of a message. As soon as it is received, its contents are analysed and a process to verify or sign the message is performed. The message is then returned to the facilities layer and the system stays again waiting for new messages. The process to create the signature generation or signature verification of a message is detailed in Fig. 2. There are two important functions defined by the IEEE 1609.2 Standard [6],

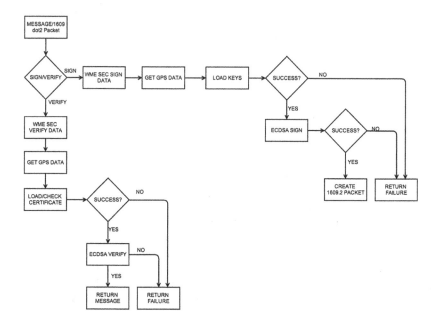

Fig. 2. Messages' signature generation and verification process

the WME_SEC_SIGN_DATA and the WME_SEC_VERIFY_DATA, which are responsible to handle the protocols to sign or verify the messages correctly. These functions access the GPS data and retrieve the keys.

3.2 ECDSA Implementation

The ECDSA algorithm was implemented in C programming language using the OpenSSL library.

Open-SSL API: In this section, the OpenSSL library which performs the required cryptographic algorithm with the specific curves is explained.

Key Generation: The key generation process is not ECDSA specific and it is generated in the following way [4]:

Algorithm 1. ECDSA Key generation

```
EC_KEY *eckey = EC_KEY_new;
if (eckey == NULL) then
  {Handle Error}
else
  EC_GROUP *ecgroup = EC_GROUP_new_by_curve_name(int nid);
  if ecgroup == NULL then
    {Handle Error}
  else
    int set_group_status=EC_KEY_set_group(eckey,ecgroup); /*Return 1 for suc-
    cess*/
    if set_group_status != 1 then
      {Handle Error}
    else
      int gen_status = EC_KEY_generate_key(eckey); /* Return 1 for success*/
      if gen_status != 1 then
        {Handle Error}
      end if
    end if
  end if
end if
```

The NID value that is assigned in the function EC_GROUP is the name of the curve that is used to create the Elliptic Curve Group, which in this case are the NIST prime curves defined respectively by the following names: NID secp224r1 and NID secp256k1 [4]. The EC_KEY *eckey is a structure that is composed of parameters that define the type of key generated along with the private and public key. It's structure is defined bellow, and when the EC_KEY_new(*void) is called; it is created the following way:

- eckey→ version = 1;
- eckey → group=NULL;
- eckey → pub_key=NULL;
- eckey → priv_key=NULL;
- eckey → enc_flag=0;

- eckey → references=1;
- eckey → method_data=NULL;
- eckey → conv_form=POINT
  _CONVERSION_UNCOMPRESSED;

Signature Generation: For the signature generation, considering that the private key is already generated, and apart from the data processing, it is only needed to call the following Open-SSL function:

Algorithm 2. ECDSA Signature Generation

ECDSA_SIG * = ECDSA_do_sign(const unsigned char *dgst, int dgst_len, EC_KEY *eckey);

This function has as return value the signature that was generated for the input data, with twice the size of the key length used in the ECDSA algorithm.

Signature Verification: The verification of a signature is composed of several steps that must be taken into consideration to make a verification with OpenSSL:

Algorithm 3. ECDSA Signature Verification

EVP_PKEY pk = EVP_PKEY_new();
EC_KEY publickey;
pk = X509_get_pubkey(X509);
publickey = EVP_PKEY_get1_EC_KEY(pk);
int ECDSA_do_verify(const unsigned char dgst, int dgst_len, const ECDSA_SIG sig, EC_KEY eckey);
EVP_PKEY_free(pk);
EC_KEY_free(publickey);

The return value from the ECDSA_do_verify function determines if the verification was a success (return value = 1), failure (return value = 0) or an error occurred (return value = -1).

Certificates: The IEEE 1609.2 Standard defines the WAVE-Certificates which are a special type of certificates for Vehicular Communications. As PKI was not implemented, the standard certificates generated using OpenSSL were used. The used certificates are encoded in a specific format (X.509) and their size can vary, depending on the key length used for the public key algorithm, on the size of identification and on some optional values. In this implementation, the certificates had approximately 956-bytes which are too big when compared to the expected WAVE Certificate size of about 120-bytes [5]. An application to generate private and public keys together with the respective certificate was implemented separately to support the main application that makes use of this cryptographic material.

3.3   Implementation of Secure Protocols

So far the cryptographic engine supported by OpenSSL libraries is able to perform all the required algorithms from the standard. As already mentioned, a cryptographic engine is not enough if it is not strongly supported by interface protocols. The IEEE 1609.2 D17 Draft Standard defines these protocols to use with the ECDSA algorithm. These protocols are handled by the Wave Management Entity (WME), which is responsible for the handling of data from higher layers. The WME also makes requests to the security module to ask for secure data. Secured data is sent over the network within a secure packet structure defined by IEEE 1609.2. Its structure is composed of multiple sub-structures as it is shown in Fig. 3. Figure 4 illustrates the implemented protocols for the signature generation and verification.

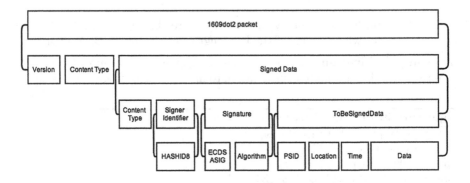

Fig. 3. 1609.2 secure packet struct

4   Integration with WSMP and Facilities Layer

After the security services have been developed there was the need to integrate this work with other applications such as the WAVE Short Message Protocol (WSMP) and the facilities layer in order to have a full system working. Sockets were used to interconnect the different applications and create an architecture capable of generating messages, secure them and communicate through the Dedicated Short Range Communications (DSRC) platform IT2S developed in the IT (Telecommunications Institute) in the scope of the FP7 project ICSI.

A Cooperative Awareness Message (CAM) message is generated by the facilities module with all the information gathered from the vehicle: speed, location, direction and all types of valuable information. This message is sent to the security services which receives the message and digitally signs its content with its private-key. After the message gets secured in the format of a 1609.2 packet, it is sent back to the facilities module which forwards its content to the WSMP. The WSMP generates the WSM packet and puts in its data field the 1609.2 packet received from the facilities module. This packet is then transmitted over the DSRC platform.

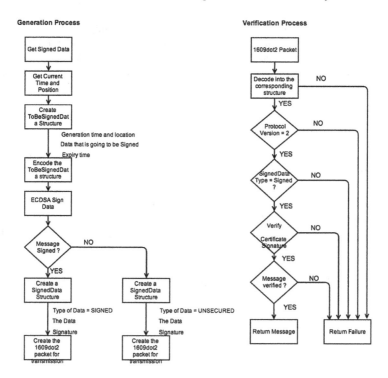

Fig. 4. Signature generation and verification protocol

Table 1. Flow detection and mitigation time

Specification	Personal laptop	Raspberry-Pi
CPU	Intl. Pentium 1.46 GHz	700 MHz Low Power
Instruction-Set	32-bits	32-bits
Memory	3 GB	512 MB SDRAM
OS	Ubuntu 12.04	Arch-Linux

5   Experimental Evaluation

This section presents a experimental performance analysis of the implemented architecture. The main focus of these experiments was to benchmark the overall system that provides authenticity with ECDSA. In the first approach, the system was benchmarked in a laptop and later in a Raspberry-Pi which was the option as the on-board computer for a real vehicular communication system. The choice to use two computers in the experiments was to clearly understand how much computational power might be needed to achieve a good performance of the system. Table 1 provides hardware and software comparison between the two used machines. The following set of experiments were carried out:

- ECDSA algorithm timing analysis on Laptop with increasing random payload;
- ECDSA algorithm timing analysis on Raspberry-Pi with increasing random payload;
- Security Implementation timing analysis on Laptop with CAM Messages as payload;
- Security Implementation timing analysis on Raspberry-Pi with CAM Messages as payload.

5.1   ECDSA Timing Performance Analysis

For the analysis of the execution times regarding the ECDSA algorithm, an application capable of signing and verifying messages without all the overhead caused by the security services was developed. In order to calculate the execution time of the algorithm to be evaluated, a processor tick timer was inserted in the code to count the number of ticks that each function runs. Then the execution of the code was analysed for the ECDSA NIST curve P224 and P256 with the payload varying from 10-bytes to 2000-bytes. Within each payload, the code runs 1000 times in order to calculate the mean execution time for the given payload size. The payload was randomly generated, containing only alpha-numeric values. In Table 2 summary of the mean execution time for both key algorithms and computers is presented.

Table 2. Experiment results for ECDSA 224 and 256

Computer	Algorithm	Sign [ms]	Signature/s	Verify [ms]	Verifications/s
Laptop	ECDSA P256	2.339	411	2.8676	349
Laptop	ECDSA P224	1.8958	527	2.222	450
Raspberry-Pi	ECDSA P256	11.3833	88	13.3581	75
Raspberry-Pi	ECDSA P224	8.7552	114	10.2238	98

It is important to mention that all these values only represent timings of the OpenSSL signature generation and verification functions. For the OpenSSL functions to sign and verify, an ECDSA signature or verification requires that a key-pair must be previously generated or loaded and a hash function must be applied to the payload. These values do not represent the overall system execution times but they were taken to benchmark the OpenSSL library.

5.2   Integration of CAM, WSMP and Security

Here we present an analysis of timings with integration of WSMP, the facilities application and security. This step is important to better understand how the system handles the increase payload, the time for the whole process, the signature generation, the signature verification, loading the certificates and the hashing of data. We used a real testbed having all the defined modules integrated and transmitting the CAM messages at the rate of 10 Hz. Ten thousand messages were sent at this rate to analyse the system performance. Two different size of messages were generated, one with 53 bytes and another one with 67 bytes. In Fig. 5, a comparison between signing and verifying for both messages is presented. The mean time to sign and to verify a message regardless of the size (53 or 67 bytes) is 3:8504 ms and 4:4157 ms respectively. Figure 5 shows that the times to perform a signature generation and verification are very high with the following mean times: signature generation: 21:7615 ms and signature verification: 25:3628 ms, considering both the 53 and 67 byte payload of the CAM message (Fig. 6).

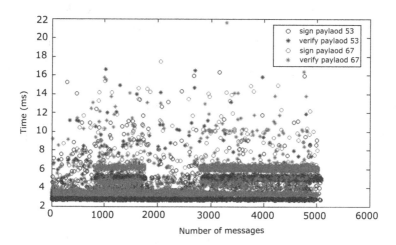

Fig. 5. CAM timings obtained on laptop

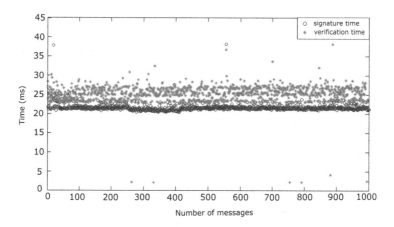

Fig. 6. CAM timings obtained on Raspberry Pi

6   Conclusions

The main focus of this work was to analyse the requirements and how much effort has to be applied to provide the adequate security requirements for a vehicular network. The proposed software implementation was based on the IEEE 1609.2 TM/D17 standard and the entire system was evaluated in a real world scenario. Performance tests show that the system is capable of performing 223 and 39 signature verifications per second when running on a conventional laptop or Raspberry-Pi respectively. Since in WAVE, cars are typically beaconing at 10 Hz, the maximum number of cars in the neighbourhood possible to verify are 22 using laptop and 4 using RaspBerry-Pi respectively. However in real world scenarios, specially in highway and congestion scenarios, the number of vehicles in a given area can reach to hundreds. Thus, it can be concluded that the performance of a pure software implementation of security services is insufficient for real world use-cases. As future work, the study of other algorithms besides the ones referred in the IEEE 1609.2 TM/D17 Standard should be analysed. Also a hardware solution based on FPGA could be developed to perform some of the arithmetic operations required by the cryptographic algorithms. The hardware module can be integrated with the OpenSSL library allowing some hard operations to be performed on hardware.

Acknowledgment. This work was funded by the European Union's Seventh Framework Programme (FP7) under grant agreement no. 3176711.

References

1. Sousanis, J.: WardsAuto. World Vehicle Population Tops 1 Billion Units, August 2011. http://wardsauto.com/news-analysis/world-vehicle-populationtops-1-billion-units
2. Alam, M., Ferreira, J., Fonseca, J.: Introduction to intelligent transportation systems. In: Alam, M., Ferreira, J., Fonseca, J. (eds.) Intelligent Transportation Systems. SSDC, vol. 52, pp. 1–17. Springer, Heidelberg (2016). doi:10.1007/978-3-319-28183-4_1
3. Alam, M., Fernandes, B., Silva, L., Khan, A., Ferreira, J.: Implementation and analysis of traffic safety protocols based on ETSI Standard. In: Proceedings of the 2015 IEEE Vehicular Networking Conference (VNC), pp. 143–150, Kyoto (2015)
4. Hudson, T., Young, E.: OpenSSL - Cryptography and SSL/TLS Toolkit. https://www.openssl.org/. Accessed 30 June 2016
5. Hartenstein, H., Laberteaux, K.: VANET Vehicular Applications Inter-Networking Technologies. Wiley, Hoboken (2010). ISBN: 978-0-470-74056-95
6. IEEE Standard for Wireless Access in Vehicular Environments Security Services for Applications and Management Messages. doi:10.1109/ieeestd.2013.6509896
7. Papadimitratos, P., et al.: Secure vehicular communication systems: design and architecture. IEEE Commun. Mag. **46**(11), 100–109 (2008). doi:10.1109/mcom.2008.4689252
8. Alam, M., Sher, M., Afaq Husain, S.: VANETs mobility model entities and its impact. In: 2008 4th International Conference on Emerging Technologies, October 2008
9. Alam, M., Fernandes, B., Silva, L., Khan, A., Ferreira, J.: Implementation and analysis of traffic safety protocols based on ETSI Standard. In: Proceedings of the 2015 IEEE Vehicular Networking Conference (VNC), December 2015

Performance Evaluation of SIMO Techniques in IEEE 802.11p

Nelson Cardoso[1], Muhammad Alam[1], João Almeida[1,2(✉)], Joaquim Ferreira[1,3], and Arnaldo S.R. Oliveira[1,2]

[1] Instituto de Telecomunicações, Aveiro, Portugal
{nelsoncardoso,alam}@av.it.pt
[2] DETI, Universidade de Aveiro, Aveiro, Portugal
{jmpa,arnaldo.oliveira}@ua.pt
[3] ESTGA, Universidade de Aveiro, Águeda, Portugal
jjcf@ua.pt

Abstract. In this paper, we evaluate the performance of spatial diversity techniques in the scope of vehicular communications, using a IEEE 802.11p multi-antenna flexible and modular implementation. The analysis is based on the well-known Selection Combining (SC) and Equal Gain Combining (EGC) algorithms. The results obtained indicate that employing these techniques improves the system's resilience against the fast fading multipath effect that characterizes dynamic wireless channels. The measurements performed reveal that up to 32% more frames can be decoded when compared to a single antenna system within a Non Line of Sight (NLOS) scenario. We also examine the influence of the modulation type used on the performance of the diversity combining scheme.

Keywords: SIMO · IEEE 802.11p · Vehicular communications

1 Introduction

Vehicular communications have raised interest in the last few years and are currently under intense research and development worldwide. They follow the Dedicated Short Range Communication (DSRC) paradigm and enable a wide range of applications for Intelligent Transportation Systems (ITS), such as road safety, traffic efficiency, real-time traffic information, entertainment and comfort [1]. The communication between wireless nodes in a ITS scenario could be either vehicle to infrastructure (V2I) [2] or vehicle-to-vehicle (V2V) and are based on the IEEE 802.11:2012 - amendment 6, also know as IEEE 802.11p standard [3] for Wireless Access in Vehicular Environments (WAVE). In the United States a 75 MHz bandwidth has been allocated for DSRC in the range from 5.850 to 5.925 GHz, whereas in the European Union 50 MHz has been allocated ranging from 5.855 to 5.905 GHz.

The vehicular environment is highly dynamic and complex due to the large number of moving vehicles and the fast change of the surrounding area. This

© ICST Institute for Computer Sciences, Social Informatics and Telecommunications Engineering 2017
J. Ferreira and M. Alam (Eds.): Future 5V 2016, LNICST 185, pp. 91–100, 2017.
DOI: 10.1007/978-3-319-51207-5_9

leads to a situation where Radio Frequency (RF) signals can bounce off several obstacles. This scattering implies that many copies of the signal may arrive to the receiver, having travelled along many different paths. When these copies combine, they may add constructively, giving a good overall signal, or destructively, mostly canceling the received signal. This fast fading multi-path effect of the channel is highly frequency selective. This means that some frequencies in a 10 MHz 802.11p channel may be wiped out while others are unaffected. On the other hand the typical vehicular communication suffers from high Doppler shifts due to the relative movement of the nodes [4].

One of the main purposes of vehicular communications is to support safety-related applications. Typically these services demand a robust, very low latency and high-reliable communication system between the involved nodes. This could be a challenge taking into account the problems already mentioned. One way to overcome the fast fading multi-path effect, without violating the standard, is the use of multiple antennas at the receiver, improving the system reliability. In this paper we focus on the study and evaluation of diversity techniques and its impact on communication reliability using the IT2S IEEE 802.11p platform. IT2S is a modular, flexible and multiradio FPGA (Field-programmable Gate Array) based platform designed at IT-Aveiro for research and development on vehicular communications. The flexibility of the IT2S platform, permitted by its modular design, consisting in independent blocks for the RF, baseband PHY and upper layers software, allow us to easily modify the Physical Layer (PHY) in order to include several diversity methods in the reception side. In the context of this work we evaluate the performance of Selection Combining (SC) and Equal Gain Combining (EGC) versus single antenna based on laboratory measurements.

The reminder of this paper is structured as follows: the next section gives a short overview of relevant related work. Section 3 describes the diversity methods used in this work. The technical implementation is described in Sect. 4. Section 5 gives an overview concerning the testbed setup. The measurements results are presented in Sect. 6. In the last section we draw the conclusions and some outlines for future work.

2   Related Work

Nuckelt et al. [4] performed a series of simulations based on several V2V and V2I scenarios. They applied three different diversity techniques: SC, EGC and Maximum Ratio Combining (MRC) with up to four antennas at the receiver. The channel and the 802.11p PHY layer were both modeled in MATLAB. The obtained results shown a slightly improvement for SC and significant increased performance for EGC and MRC with MRC outperforming all other combining techniques. However, the study performed in this work is based on simulations with short packets of 100 bytes, limiting its generality and scope. It would be interesting to analyze the system's performance with bigger frame sizes and different modulations schemes, as well as performing the validation of the results in practice, in order to evaluate the techniques in a real world scenarios.

Real-World measurements can be found in [5,6], where the authors carried out some measurement campaigns with real V2V and V2I scenarios. Several different approaches for diversity schemes have been applied in this work, particularly SC, EGC and MRC. The analysis of the results reveals a better performance of EGC and MRC when compared to SC. Although in [5] the performance of EGC and MRC was very similar. The technical solution implemented for the diversity schemes is very similar to the one described in this paper. However, it yields a very complex and bulky hardware support when compared to our own hardware platforms. On the other hand, in [5], only the automatic gain control (AGC), time and frequency synchronization were implemented in the FPGA. The rest of the forward processing modules were programmed in a digital signal processor (DSP) which results in a high latency system. In contrast, our solution includes all the processing modules that are completely implemented in the FPGA, thus reducing the overall system's latency.

In this work, we present a different approach by applying receive diversity techniques directly to the IT2S modular and compact IEEE 802.11p multi-radio platform which has been adapted for that purpose. Starting from a complete PHY implementation for single antenna we were able to perform the modifications needed in order to achieve receive diversity for different data rates. In fact, the IT2S multi-radio capabilities allow its use in different communication scenarios, namely in simultaneous multichannel communications (as defined in ETSI ITS G5 standards) using single antenna and independent PHYs, as well as in the exploration of diversity techniques using multiple antenna configurations, as discussed in this paper. These capabilities make the IT2S platform adaptable to different communication scenarios depending on the required throughput, signal strength, distance range, etc.

3   Overview of the Diversity Techniques

In this section we provide a short overview about the applied receive diversity schemes.

The main goal of receiving antenna diversity is to take advantage of multiple versions of the transmitted signal that travelled along independently faded paths, and combine them into a single improved signal. Typically we can find in the literature [7] three methods for diversity: spatial, time and frequency diversity. Within the scope of this analysis, we focus on signal diversity obtained by spatial separation of the receiver antennas. So a system called Single-Input Multiple-Output (SIMO) is used to achieve spatial diversity.

The simplest diversity method commonly used is to choose the receiver's antenna with the strongest signal. This method, called Selection Combining (SC), improves reliability, because both signals are unlikely to be simultaneously weak. It is also easily implemented in hardware due to its low complexity. However, it wastes the received power at the antenna that was not selected. This particular method should not be considered as a true SIMO system because, in fact, only one antenna is used each time a received frame is processed. A potential better method uses the incoming independently faded signals from both antennas in order to combine them coherently, resulting in a single signal with improved Signal to-Noise Ratio (SNR). According to [6] a linear diversity combiner with two antennas could be defined by

$$Y_{n,k} = \alpha y_{n,k}^{(ch0)} + \beta y_{n,k}^{(ch1)} \tag{1}$$

where α and β are the weighting factors, which in our case are set to one thus leading to a method called Equal Gain Combining (EGC). $y_{n,k}^{(i)}$ represents the received data at antenna (i) at time n and subcarrier index k. This technique requires two dedicated RF frontends for each antenna which increases the hardware complexity and power consumption but yields better performance than SC.

4   Implementation of the Diversity Techniques

In this section, we give a detailed description on the system architecture implemented in a FPGA for both diversity schemes evaluated.

4.1   Selection Combining

The Fig. 1 depicts the complete block diagram for the SC method solution. All the blocks represented are implemented in the FPGA except the RF frontends. This solution is a quick and simplified adaptation of the existing reception (RX) chain for single antenna. Obviously this model can be further optimized in terms of resources, but for a simple proof of concept it works perfectly well for the SC analysis. The incoming frame is initially processed by both RF front ends. During the firsts short training sequences of the preamble, the Received Signal Strength Indication (RSSI) is measured for each branch and fed to the corresponding AGC

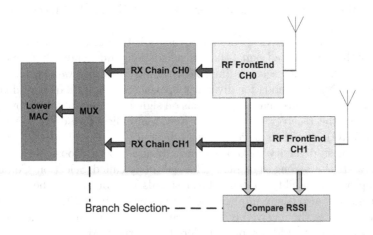

Fig. 1. Selection combining block diagram.

module. At the same time a comparator checks both RSSI values and chooses the higher one. Finally the output of the comparator forces the multiplexer to link the selected RX chain branch with the Lower MAC block for further frame processing. This simple process is repeated for each received frame.

4.2 Equal Gain Combining

The Fig. 2 shows the block diagram of the RX chain used for the EGC technique with two antennas. All the blocks included in this diagram are implemented in the FPGA. Once again this solution is a modification of the existing RX chain for a single antenna. In the first block time and frequency synchronization is performed individually by both branches. Then the received time-domain signal is transformed back to the frequency-domain using a 64-point Fast Fourier Transform (FFT). At this point the resulting samples of each branch are delayed until they are perfectly aligned with respect to the start of an OFDM symbol. The equalizer performs the channel estimation based in the long training sequences in order to determine the equalization factors: phase and amplitude. Then the subcarrier samples are phase compensated [5] by the equalizer in both branches, according to

$$\hat{y}_{n,k}^{(ch0)} = y_{n,k}^{(ch0)} exp(-jargH_k^{(ch0)}) \tag{2}$$

$$\hat{y}_{n,k}^{(ch1)} = y_{n,k}^{(ch1)} exp(-jargH_k^{(ch1)}) \tag{3}$$

where $H_k^{(i)}$ represent the equalization factors at antenna (i) and subcarrier index k. This operation ensures that the samples are suited to be combined coherently. Note that the amplitude compensation is only performed in the next step.

The following block is the Diversity Combiner. This module combines the two branches based on

$$\hat{\mathcal{Y}}_{n,k} = (|H_k^{(ch0)}| + |H_k^{(ch1)}|)(\hat{y}_{n,k}^{(ch0)} + \hat{y}_{n,k}^{(ch1)}) \tag{4}$$

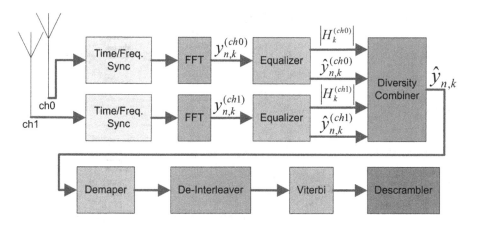

Fig. 2. Equal Gain Combining block diagram.

resulting in a single set of sub-carrier samples per OFDM symbol. The Demaper converts the I/Q vectors back to a bit stream. After de-interleaving, the binary sequence is decoded by a forward error correction block based on the Viterbi's algorithm. Finally the data is descrambled in order to retrieve the original frame.

5   Experimental Setup

In this section we present a summary of the testbed environment. The IEEE 802.11p hardware platforms used in our measurements includes two 5.9 GHz custom made RF frontends, two 10 bit AD/DA (Analog to Digital and Digital to Analog converter) processors and a FPGA module fitted in a Xilinx Spartan-6 XCSLX150-2CSG484C. The digital baseband PHY was modelled in VHDL (VHSIC Hardware Description Language) and implemented in the FPGA. The platforms were totally developed from scratch and are compliant with the 802.11p standard. We have full access to both analog and digital PHY layers, which represents a great advantage and flexibility over closed black box commercial solutions.

The Fig. 3 depicts the block diagram of the used multiradio platform. In this architecture the PHY layer is divided in two sub-layers: digital PHY and analog PHY. In this context the digital PHY sub-layer is implemented in the FPGA module. Apart from processing the transmitted and received data frames, the FPGA module is also responsible for managing the configuration, control and communication of both AD/DA processors and RF modules. It also provides interface with the GPS (Global Positioning System) module and includes a Universal Serial Bus (USB) interface for connecting the upper layers running in a single board computer. The GPS module provides location and time synchronization.

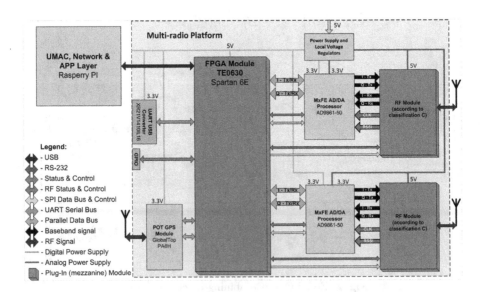

Fig. 3. Complete block diagram of the multi-radio hardware platform.

The AD/DA processor is an integrated converter and defines the interface between the digital and analog PHY sub layers. It is responsible for converting the OFDM modulated baseband signals from the digital domain to the analog domain (and vice versa). This processor also includes two auxiliary ADC (Analog to Digital Converter) channels in order to sample signals such as RSSI or the Power Amplifier (PA) emitted power (both signals generated by the RF module) for monitoring and control purposes. The RF module implements the analog PHY sub-layer according to the IEEE 802.11p standard for the 5.9 GHz class C wireless communication band. The module's main task aims to convert the I/Q components of the OFDM signals from baseband to radio frequency (and vice versa) and perform its transmission (or reception) to the medium. This custom module incorporates all the key components (transceiver, filters, PA, RF switch, etc.) for proper operation over the 5.9 GHz range. For these measurements and tests we used two hardware platforms, each one connected to a Raspberry PI via USB. The Raspberry PI is a small single board computer running one application with configurable parameters for transmission and the other one was programmed to act as a sniffer, collecting the detected and correctly decoded 802.11p packets. The transmitter was equipped with only one RF module while the receiver included two RF modules for diversity tests. In the laboratory room the distance between transmitter and receiver was about 7 m. The transmitter's antenna was placed right behind a set of metal cabinets, in a configuration of Non Line of Sight (NLOS) relatively to the position of the receiver's antennas (refer to Fig. 4). This setup creates a richly environment of scattered replicas of the transmitted signal which can be exploited by the diversity techniques. The set of antennas used were monopoles suited for the 5.9 GHz operation band. The distance between the two receiver's antennas was about 50 cm. Furthermore the equipment was tuned in the 172 channel which corresponds to a center frequency of 5.86 GHz. For all the measurements the constant transmitted power programmed was about 3 dBm. The transmit power was calibrated in order to obtain a Packet Error Ratio (PER) close to 50 single antenna in NLOS configuration. Thus, in this mode, we can more easily highlight the performance of the spatial diversity schemes under test in this work. At each transmission trial we

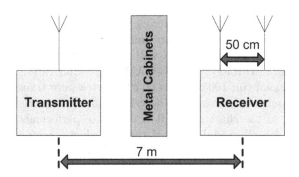

Fig. 4. Testbed setup layout.

sent a thousand packets of 1004 bytes, including the SIGNAL field, SERVICE bits, MAC header and payload. At the receiver side we measured the number of successfully decoded frames per trial.

6   Experimental Results

The Fig. 5 represents the number of successfully decoded packets obtained for a single antenna receiver within NLOS and Line of Sight (LOS) regimes. In this particular test we sent 1000 frames at 6 Mbits/s for different transmitted power values in a laboratory environment. The results for LOS regime indicate a perfect reception rate, except for slightly performance decrease at -3 dBm resulting in a PER of 4.8%. On the other hand in the NLOS regime we can observe a perfect reception rate above 9.8 dBm. Bellow this value there is an important performance degradation in terms of received frames. This is precisely the point were we can enable and take advantage of the receive diversity techniques.

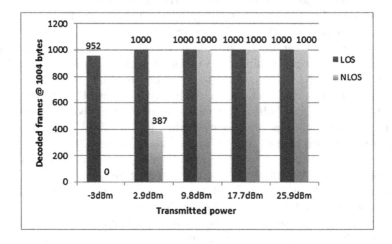

Fig. 5. Single antenna measurements results.

The values in Fig. 5 depicts the average value of 10 measurement runs performed for each parameter setting. Three modulation schemes have been chosen for this performance evaluation: BPSK with coding rate 1/2 (3 Mbits/s), QPSK with coding rate 1/2 (6 Mbits/s) and 16-QAM with coding rate 1/2 (12 Mbits/s). For each transmission run 1000 frames of 1004 bytes were transmitted and the corresponding number of successfully decoded frames at the receiver was measured. According to the chart of Fig. 6, a moderate performance improvement can be observed for all modulations of the SC scheme compared to the single antenna configuration (named antenna ch0). In fact, we observed during the tests that the SC system often opted by the antenna ch1, discarding the signal at antenna ch0. This behavior clearly shows that antenna ch1 constantly picked-up a stronger signal than antenna ch0, within this particular testbed configuration.

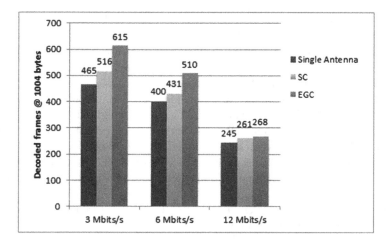

Fig. 6. Multiple antenna measurements results.

In the best scenario, which corresponds to 3 Mbits/s, the SC technique was able to decode up to 11% more frames than the single antenna setup. Results for EGC scheme indicate a significantly increased performance, especially for 3 and 6 Mbits/s, when compared with the single antenna. This is an expected result because the EGC technique takes advantage of the signals from both antennas in order to combine them into a single signal with improved SNR, thus leading to an increased number of successfully decoded frames. In this case the EGC scheme was able to decode up to 32% more frames when compared to the single antenna results for 3 Mbits/s.

Another interesting observation on the results is that the overall number of decoded frames decreases with the increase of the data rate. In particular at 12 Mbits/s (16-QAM) which denotes only a slightly performance improvement for SC and EGC modes. This can be explained by the fact that higher modulations are more sensitive to the fast fading multi-path effect in NLOS configuration resulting in a higher Error Vector Magnitude (EVM) that cannot be significantly compensated by the diversity techniques employed in this work. So, in this testbed setup, SC and EGC performed better at lower modulations schemes.

7   Conclusions and Future Work

In this paper we presented the evaluation of two different spatial diversity combining techniques. SC and EGC were implemented in our IEEE 802.11p IT2S platforms. Several laboratory measurements have been carried out in order to analyze the resulting performance gain of these techniques when compared with the single antenna setup. The presented results show that the system performance is improved if SC or EGC are applied. More specifically we have shown that with the use of SC, up to 11% more packets can be decoded. EGC performed even better, with up to 32% more frames decoded compared to a single antenna for 3 Mbits/s, resulting in a more robust and reliable communication system. We also shown that the performance of the spatial diversity schemes applied depends on the type of modulation used.

For future work we will take into consideration the use of MRC diversity technique, where the factors α and β (in Eq. 1) are weighted according to the SNR estimated for each branch. We also intend to carry out some real world experiments with moving vehicles, equipped with our platforms, at different speeds within V2V and V2I scenarios. We will consider several environments such as open space (LOS), tunnel and city with different traffic conditions.

Acknowledgements. This work is funded by National Funds through FCT - Fundação para a Ciência e a Tecnologia under the PhD scholarship ref. SFRH/BD/52591/2014 and the project PEst-OE/EEI/LA0008/2013.

References

1. Alam, M., Ferreira, J., Fonseca, J.: Introduction to intelligent transportation systems. In: Alam, M., Ferreira, J., Fonseca, J. (eds.) Intelligent Transportation Systems. SSDC, vol. 52, pp. 1–17. Springer, Heidelberg (2016). doi:10.1007/978-3-319-28183-4_1
2. Alam, M., Fernandes, B., Silva, L., Khan A., Ferreira, J.: Implementation and analysis of traffic safety protocols based on ETSI standard. In: Proceedings of the 2015 IEEE Vehicular Networking Conference (VNC), Kyoto, pp. 143–150 (2015)
3. IEEE Standard for Information technology-Telecommunications, information exchange between systems Local, metropolitan area networks-Specific requirements Part 11: Wireless LAN Medium Access Control (MAC) and Physical Layer (PHY) Specifications, pp. 1583–1630 (2012)
4. Nuckelt, J., Hoffmann, H., Schack, M., Kurner, T.: Linear diversity combining techniques employed in Car-to-X communication systems. In: Proceedings of the IEEE 73rd Vehicular Technology Conference (VTC Spring), pp. 1–5 (2011)
5. Maier, G., Paier, A., Mecklenbrauker, C.: Packet detection and frequency synchronization with antenna diversity for IEEE 802.11p based on real-world measurements. In: International ITG Workshop on Smart Antennas (WSA), pp. 1–7 (2011)
6. Maier, A., Paier, G., Mecklenbrauker, C.: Performance evaluation of IEEE 802.11p infrastructure-to-vehicle real-world measurements with receive diversity. In: 8th International Wireless Communications and Mobile Computing Conference (IWCMC), pp. 1113–1118 (2012)
7. Lozano, A., Jindal, N.: Transmit diversity vs. spatial multiplexing in modern MIMO systems. IEEE Trans. Wireless Commun. **9**(1), 186–197 (2010)

Wireless Power Transfer for Energy-Efficient Electric Vehicles

Wael Dghais[1](⊠) and Muhammad Alam[2]

[1] Department of Microelectronic, Institut Supérieur des Sciences Appliquées et de Technologie de Sousse, Université de Sousse, Sousse, Tunisia
wael.dghais@hotmail.co.uk
[2] Instituto de Telecomunicações, Aveiro, Portugal
alam@av.it.pt

Abstract. This paper presents the wireless power transfer (WPT) technology based on inductive coupling and the design challenges of a hybrid energy harvester (EH) circuit as a promising solution to promote the energy efficiency of the electric vehicles (EVs). The design methodologies of ultra-low power electronic module based on low leakage conditioning and processing device are details based on nanoscale transistor technology so that the WPT and hybrid EH can be implemented for self-powered devices in EVs.

Keywords: Adaptive back-gate biasing · Electric vehicles · Energy harvester · Ultralow power design · Wireless power transfer

1 Introduction

The share of electronic component by the high integration of more sensors and actuators in the EVs is fast growing and plays a decisive role not only in satisfying primary customer wishes for better driving safety (vehicle stability in windy or rainy environment) and comfort such as entertainment applications (i.e. music, video), but at the same time to achieve better electric energy economy [1]. Moreover, the contribution of numerous digital, analog, and mixed signal processing transceivers as well as the high-integration of electronic control units (ECUs), high definition screens makes the embedded devices much more energy hungry in EVs [2–4]. Many of these functions has to be designed and implemented considering the wireless data exchange between electronic components and the optimizing the energy consumption to prolong the EV's battery autonomy.

The wireless power transfer (WPT) technology and energy harvesting present the effective solution to energy-efficient EVs to stay competitive in the market share [2–5]. The WPT based on inductive coupling mechanisms and the design principles for green EVs based on self-powered micro-electro-mechanical systems (MEMS) and their implementation in a most cost-effective way play an important role in order to improve energy efficiency and reduce greenhouse gas emissions in the sustainable transport sector [4–6]. The remains of this manuscript is organized as follows. Section 2 describes the inductive coupling mechanism and its performance in promoting not only

© ICST Institute for Computer Sciences, Social Informatics and Telecommunications Engineering 2017
J. Ferreira and M. Alam (Eds.): Future 5V 2016, LNICST 185, pp. 101–111, 2017.
DOI: 10.1007/978-3-319-51207-5_10

the WPT for powering the EV's sensor or actuator but also communication between vehicle's wireless sensor networks. Section 3 presents low static power design techniques through back-gate biasing design methodologies enabled by the new nanoscale transistor technology without sacrificing chip's speed while detailing the back-gate biasing mechanisms and effects and presenting the fundamental techniques to reduce leakage power. Finally, conclusions are drawn in Sect. 4.

2   Wireless Power and Data Transmission

2.1   Inductive and Radiative Coupling

Self-powered electronic devices (i.e. processors, sensors, and actuator) are free of cables and have freedom mobility during charging and usage. The device's charging use the WPT principle which is based on a magnetic resonance coupling. Inductively coupled systems are based upon a transformer-type coupling between the primary and the secondary coils (antenna). For instance, radio frequency identification (RFID) is a wireless communication technology based on the WPT principle of mutual induction used for in-vehicle communication but can be used also for near field communication between close vehicles or for localization purpose [5] as shown in Fig. 1.

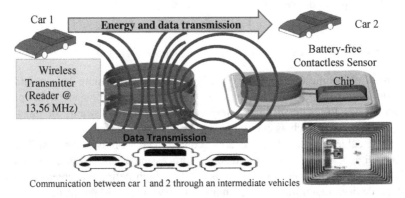

Fig. 1. Inductive coupling for wireless power and data communication between vehicles.

The RFID device of the first car is self-powered by the received electromagnetic wave and is capable of communicating data with reader of the second car that modulates the RF signal that is coming from the reader. The power consumption on the reader is minimized because there is no RF section embedded in the chip and the information is communicated through load modulations [4, 5]. The RFID and inductively coupled devices may be modelled, as a first approximation by the circuit shown in Fig. 2 where L_1 and L_2 represent the inductance of the transmitter and receiver antenna, respectively. R_1 and R_2 are the antenna's coil resistances. The current consumption of the data memory is modeled by the load resistor R_L. A time varying

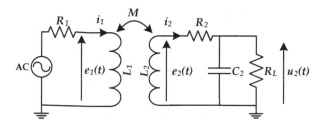

Fig. 2. Equivalent circuit diagram for magnetically coupled conductor loops

modulated magnetic field in the coil L_1 induces voltage $u_2(t)$ in the coil loop L_2 due to mutual inductance M. The flow of current creates an additional voltage drop across the coil resistance R_2.

Under sinusoidal RF approximation, the induced voltage in the receiver coil:

$$U_2(j\omega) = \frac{j\omega M I_1(j\omega)}{1 + \frac{(j\omega L_2 + R_2)}{R_L}} \tag{1}$$

The voltage $u_2(t)$ induced in the receiver coil is used to provide the power supply to the data memory (microchip). In order to significantly improve the WPT efficiency, an additional capacitor C_2 is connected in parallel with the receiver's coil L_2 to form a resonant circuit at frequency of the RFID system (i.e. $f_{RES} = 13.56\,\text{MHz}$). The required capacitance for the capacitor C_2 is found by taking into account the parasitic capacitance C_p (e.g. $C_2 = 1/\omega^2 L_2 - C_p$). Thus obtaining the relationship between voltage u_2 and the magnetic coupling of transmitter coil and transponder coil [5].

$$U_2(j\omega) = \frac{j\omega k \sqrt{L_1 L_2} I_1(j\omega)}{1 + (j\omega L_2 + R_2)\left(\frac{1}{R_L} + j\omega C_2\right)} \tag{2}$$

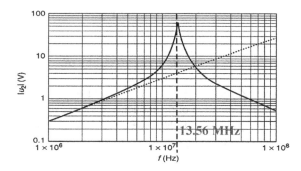

Fig. 3. Magnitude of the received voltage versus frequency. A transponder coil with a parallel capacitor (continuous line) (dashed-line: coil without a resonant capacitance).

2.2 Wireless Power Transfer

A wireless charging system enable EVs to be continuously charged with unlimited driving range. As shown in Fig. 4(a), WPT is used to wirelessly transmit large electric currents between metal coils of the high way and the EVs [2, 6]. The variable capacitor bank in the power transmitter and receiver in used to ensure the maximum WPT at the resonance frequency. Design cost to develop all electric highway will be increased with many transmitters in a given length of the track. Several transmitter coils can be connected to a single power converter in parallel. Figure 4(b) shows equivalent circuit considering various nearby transmitters which are separately powered [1].

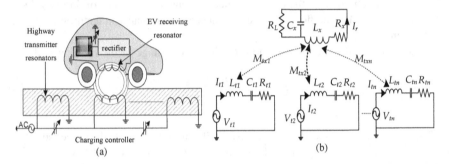

Fig. 4. (a) Magnetic fields (red) continuously and wirelessly charge EVs. (b) Equivalent circuit of EV charging considering separated powered n transmitters. (Color figure online)

Kirchhoff circuit laws allow establishing this system of linear equation that can be determined with respect to the unknown currents vector and to determine the power transfer efficiency (PTE) [1].

$$\begin{bmatrix} Z_{t_1} & 0 & \cdots & 0 & X_{t_1 r} \\ 0 & Z_{t_2} & \cdots & 0 & X_{t_2 r} \\ 0 & 0 & \ddots & Z_{t_n} & X_{t_n r} \\ X_{rt_1} & X_{rt_2} & \cdots & X_{rt_3} & (Z_r + R_L) \end{bmatrix} \begin{bmatrix} I_{t_1} \\ I_{t_2} \\ I_{t_n} \\ I_r \end{bmatrix} = \begin{bmatrix} V_{t_1} \\ V_{t_2} \\ V_{t_n} \\ 0 \end{bmatrix} \tag{3}$$

$$PTE = \frac{P_{out}}{P_{in}} = \frac{R_L |I_r|^2}{R_{t_1}|I_{t_1}|^2 + R_{t_2}|I_{t_2}|^2 + R_{t_n}|I_{t_n}|^2 + (R_r + R_L)|I_r|^2} \tag{4}$$

The self-impedances, $Z_k = R_k + j(\omega L_k - 1/\omega C_k)$, of transmitter $k \in (t_1, t_2, \ldots, t_n)$ or the receiver (r) coils are composed of the inductance, L_k, the resistance R_k of the coil, and the capacitance, C_k. $X_{jk} = X_{kj} = j\omega M_{jk}$ where $M_{jk} = M_{kj}$ is the mutual inductance between k^{th} coil and j^{th} coil, k and $j \in \{t_1, t_2, \ldots, t_n, r\}$.

2.3 Hybrid Energy Harvester

EH is an appealing and promising solution to improve the energy-efficiency of EVs. The hybrid EH focus on the electronic design of an ultra-low power (ULP) embedded devices to be powered from renewable and multiple energy sources (e.g. solar, thermal, vibration, RF signals). Firstly, ambient RF signals become widely and frequently present over an increasing range of frequencies and power levels, including mobile telephones, mobile base stations, and television/radio broadcast stations, especially in highly populated urban areas that harvest hundreds of microwatts that will potentially enable users to provide self-powered devices based in scavenged RF signals [7]. For instance, Powercast demonstrated ambient RF energy harvesting at 1.5 miles ($\sim$2.4 km) from a 5 kW AM radio station [8, 9]. While the amount of the harvested RF energy source is appropriate to bias small sensor, it is insufficient to fulfil the ECU energy requirements. Therefore, additional renewable energy in conjunction with harvested RF energy is needed to be integrated in the hybrid EH. This will lead to significantly save the electric energy and prolong the autonomy of the battery [9]. Harvesting energy from vibration use piezoelectric materials to convert mechanical strain into useable electrical energy. The piezoelectric transducer used in conjunction with a power conditioning circuit, which converts the AC output to a regulated DC output. Heat can also be easily harvested in EV. The thermoelectric generators transducer, due to the high temperature gradient and allied materials, capture waste heat in vehicles and convert it to electricity needed to power wireless sensors and transmit data or to charge batteries that run ECU.

The architecture of the adaptive multisource EH is shown in Fig. 5. This ECU of the power management module enable the multisource energy monitoring and provide control capability that will be applied to the shared maximum power point tracking (MPPT) circuit, which dynamically adjusts the operational parameters of the energy conversion devices in response to the variations of energy sources so that the output power is maximized. The rectifier circuit in the RF and vibration EH is used to convert the alternate output current at millimetre wave frequency and low frequency, respectively, into DC. In addition, a battery is used for storing a possible energy surplus.

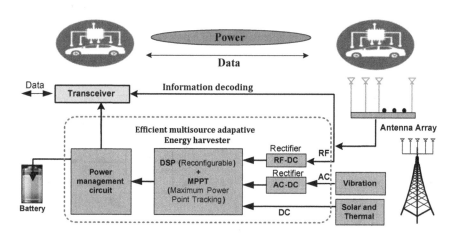

Fig. 5. Efficient Vehicle hybrid harvesting circuit

Although traditional devices are battery powered that are replaced every 3 to 5 years, electronic systems in energy-efficient EVs can be self-powered such as light adjustment systems and tire pressure monitoring systems (TPMS). For instance, MEMS-based piezoelectric EH devices is used to generate enough power (40 μW) based on the vibration's energy captured from the wheels when the car is in motion. This energy is converted to voltage that powers both the pressure sensor and the wireless communications circuitry to bias the TPMS.

Designing EVs electronic devices using ULP operation is important to prolong the battery lifetime. Legacy battery technology is finite and has not evolved at the same rate as the ultra-low power and high-speed very large scale integration for chips design, which is driven by Moore's law. Therefore, as EVs become more sophisticated, an energy trap is emerging where power demand starts to vastly exceed actual power supply. This provides an impetus for EVs stakeholders to envisage new technologies to drive future energy-efficient EVs for longer and sustainable periods of time [8].

3   Ultra-Low Power and High-Speed Chip Design

Wireless charging by means of hybrid EH and WPT requires an improved performance of low-leakage electronics design and energy storage devices in order to continuously decrease in the power consumption of electronic components. Since, the leakage current strongly depends on the threshold voltage, V_{TH}, different V_{TH} transistors can be used for speed and power tradeoff [10].

3.1   Downscaling Challenges in Bulk Technology

Transistor dimensions have been downscaled to reduce the cost, minimizing the capacitance. Moreover, the V_{dd} and V_{TH} tend to be scaled by same factor to limit current degradation. Also, the short channel effects (SCE) have a direct impact on the V_{TH} which has resulted in an exponential increase the contribution of the off-state leakage current in the total power dissipation of a bulk CMOS system as shown in Fig. 6a [11]. These consequences have moved the bulk technology to a power constrained condition.

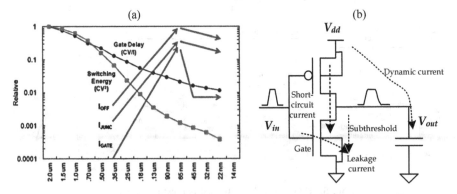

Fig. 6. (a) Ratio of active, leakage powers, and the gate delay over the CMOS technology [11]. (b) The dynamic and static (leakage) currents associated with a CMOS device.

The CMOS power consumption can be divided into three components. The dynamic and short-circuit power are consumed while the input switches. The static leakage power is consumed due to the transistor's sub-threshold, gate and diode junction's currents while the input is kept constant as shown in Fig. 6b.

$$P_{total} = \alpha \cdot C \cdot V^2 \cdot f_{clk} + V_{dd} \cdot I_{sc} + V_{dd} \cdot I_{Leakage} \tag{5}$$

The first and second terms in (5) refer to the dynamic power which represents the switching and short circuit power, P_{sw}, P_{sc}, respectively. P_{sw} is determined by the activity factor, α which is the the fraction of the circuit that is switching under the supply voltage V_{dd}, the clock speed, f_{clk}, and the equivalent switching capacitance, C. P_{sc} is consumed when both the pull up and pull down network of the logic gate circuit partially conduct as illustrated in Fig. 6b. The V_{TH} is a fundamental parameter in circuit design and testing. The transistor sub-threshold output current, I_{DS}, which is important to keep it very small in order to minimize the standby (i.e. sleep) mode. Moreover, the drain current increases exponentially on the V_{GS} [10, 12].

$$I_{DS,sub} \propto \exp\left(\frac{q.V_{GS}}{n.K.T}\right) \tag{6}$$

where K is the Boltzmann constant, T is the absolute temperature, q is the electron charge, and the sub-threshold slop n depends on the capacitance of the CMOS technology. It is worth to note that a higher I_{ON} maximizes the circuit speed because it reduces the charging time of the pad capacitances. This higher I_{ON} can be achieved by a lower V_{TH}. However, lowering V_{TH} increases exponentially the leakage current. This is the tradeoff between speed and power that the designer should balance [11, 13].

3.2 New Transistor Technology

The transistor V_{TH} can be controlled by the potential of the body terminal contact [12].

$$V_{TH} = V_{TH0} + \gamma\left(\sqrt{|-2\phi_F + V_{SB}|} - \sqrt{|2\phi_F|}\right) \tag{7}$$

where γ is the body effect coefficient, ϕ_F is the Fermi potential, and V_{TH0} is the zero threshold voltage while source-bulk bias is equal to 0 ($V_{SB} = 0$). γ describes the changes (e.g. shifting) in the V_{TH} by varying the V_{SB} voltage. It can be consider as a second gate and is sometimes referred to as the "back gate" that helps to determine how fast the transistor turns on and off. Strong γ enables a variety of effective body biasing techniques that were effectively used in older process generations. However, body effect has diminished with Bulk nanoscale transistor [12]. From transistor architecture and materials perspectives, breakthroughs were needed to reduce the SCE and the leakage currents in sub-28 nm bulk CMOS technology process and to decrease the capacitance factor. The Fin-type field-effect transistors (FinFET) and fully depleted silicon-on-insulator (FDSOI) technology provides the promising new transistor technology to do back-gate biasing effects. In addition, the SCE in an ultra-thin body

FDSOI MOSFET can be suppressed by thinning down the silicon body and buried oxide (BOX) thickness that lead to a double-gate device structure on SOI substrate. This Ultra-thin body and BOX (UTBB) FDSOI transistor architecture has a stronger γ than conventional transistors and therefore enables effective V_{TH} management through body biasing. It is worth to note also that double-gate transistor structures such as the vertical (3D) FinFET are more challenging to manufacture than the planar (2D) FD-SOI MOSFET structure as shown in Fig. 3 [12, 13]. The range of back-gate biasing in UTBB FDSOI is quite wider (i.e. $-3\,\mathrm{V} < V_{SB} < 3\,\mathrm{V}$) by a factor of 10 compared to the bulk technology (i.e. $-300\,\mathrm{mV} < V_{SB} < 300\,\mathrm{mV}$) due to the transistor structure as shown in Fig. 7. Back-biasing consists of applying a voltage just under the BOX target of the UTTB FDSOI transistors that changes the electrostatic control of the transistors and shifts their V_{TH}, as shown in Fig. 8, to speed up the switch at the expense of increased leakage current or reduce it at the expense of speed degradation [12].

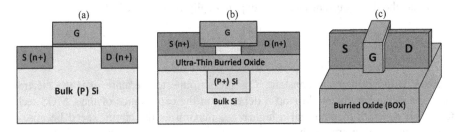

Fig. 7. Structure of different transistor technology: (a) Conventional Planar Bulk Transistor, (b) Planar Single-or double Gate FDSOI, (c) Vertical Multiple-Gate FinFET SOI [12].

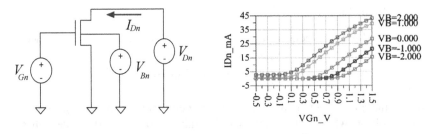

Fig. 8. Shifting effects on the V_{TH} introduced by the back-gate biasing n-channel UTTB FDSOI.

3.3   Multiple Threshold Biasing

The multiple threshold biasing technique employs the low-V_{TH} transistors to design the logic gates for which the switching speed is essential, and the high-V_{TH} transistors (also called sleep transistors) to effectively isolate the logic gates in the standby state and reduce the leakage dissipation. The generic circuit structure of the multiple threshold design circuit is offered in Fig. 9b. The sleep transistors are controlled by the sleep

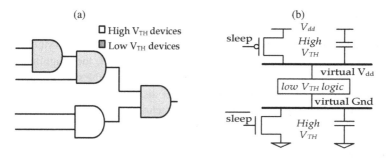

(a)
□ High V$_{TH}$ devices
■ Low V$_{TH}$ devices

(b)
V_{dd}
sleep High V_{TH}
virtual V$_{dd}$
low V$_{TH}$ logic
virtual Gnd
sleep High V_{TH}

Fig. 9. (a) Dual-V_{TH} partitioning and (b) Multiple threshold design scheme [13].

signal. During the active mode, the sleep signal is enabled, causing both high-V_{TH} transistors to turn on and provide a virtual power and ground to the low-V_{TH} logic. When the circuit is inactive, sleep signal is disabled which forces both high-V_{TH} transistors to cut-off and disconnect the power lines from the low-V_{TH} logic. This results in a very low leakage current from power to ground when the circuit is in standby mode.

3.4   Adaptive and Dynamic Back-Gate Biasing

Adaptive body bias is a valuable tool for overcoming systematic manufacturing variation, which is usually manifested in the handled devices as leakage or timing variation between chips. This undesirable current can be controlled adaptively through a body-bias circuit generator that is connected to the back-gate of the low-V_{TH} SOI nMOS and pMOS transistors as shown in Fig. 10. This dynamic control enable to dynamic shift of V_{TH} during its operation, rather than setting the body bias just once either during design or at production test, in order to either lower the V_{TH} when needing more speed, or raise it when running at lower speeds to optimize the leakage power. Consequently, dynamic body bias can be used to compensate the process variation related to the temperature, aging effects, and to efficiently manage power modes [14].

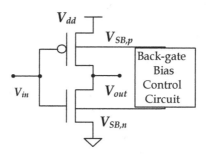

Fig. 10. Adaptive biasing scheme of low-V_{TH} and low-V_{DD} UTTB FSOI Inverter.

During the active mode the transistors circuit of Fig. 6 work as conventional CMOS transistors without back-gate biasing. As the circuit enters to the standby state, the back-gate bias control circuit generates a lower $V_{SB,n}$ for the SOI nMOS transistor and a higher $V_{SB,p}$ for the SOI pMOS transistor. As a result, the magnitudes of the respective threshold voltages $V_{TH,p}$ and $V_{TH,p}$ both increase in the standby mode due to the back-gate effect. Therefore, the leakage power dissipation in the standby state can be significantly reduced with this circuit design technique.

4   Conclusions

Highly resonant inductive coupling mechanism enable the wireless power and data connectivity between EVs that are becoming more connected and autonomous. Constructing an automated highway system and infrastructure where wirelessly EVs are charged and using a hybrid EH integrating multiple renewable energy sources will make the EVs more energy-efficient and will improve the flow of traffic by introducing more self-powered sensors and actuator while lowering greenhouse gas emissions and prolonging the battery lifetime of EVs.

Competitive energy-efficient EVs requires an improved performance of low-leakage electronics design in order to continuously decrease in the power consumption of electronic components. Therefore, the recent progress in nanoscale technology by investigating the FinFET and the UTBB FDSOI revive the ability of higher back-gate bias effect by enabling wider range of back voltage to adjust the V_{TH} according to the circuit specifications. Also, it brings a significant improvement in terms of speed, dynamic power saving and flexibility to static leakage power management design techniques for energy efficiency optimization during early silicon stage design or at the post-silicon stage by tuning the chip's bias for process compensation.

References

1. Vilathgamuwa, D.M., Sampath, J.P.K.: Wireless Power Transfer (WPT) for Electric Vehicles (EVs)-present and future trends. In: Rajakaruna, S., Shahnia, F., Ghosh, A. (eds.) Plug in Electric Vehicles in Smart Grids, pp. 33–60. Springer, Singapore (2015)
2. Alam, M., Ferreira, J., Fonseca, J.: Intelligent Transportation Systems: Dependable Vehicular Communications for Improved Road Safety, vol. 52. Springer, Switzerland (2016). ISSN 2198-4128
3. Gursoy, M., Jahn, S., Deutschmann, B., Pelz, G.: Analysis of conducted emissions in CAN bus systems with VHDL-AMS. In: EMC Europe, pp. 147–152 (2008)
4. Lu, X., Wang, P., Niyato, D., Kim, D., Han, Z.: Wireless charging technologies: fundamentals, standards, and network applications. In: IEEE Communications Surveys Tutorials, p. 9 (2015)
5. Finkenzeller, K.: RFID Handbook: Radio-Frequency Identification Fundamentals and Applications. Wiley, Hoboken (2004)
6. Miller, J.M.: Demonstrating dynamic wireless charging of an electric vehicle: the benefit of electrochemical capacitor smoothing. IEEE Power Electron. Mag. **1**, 12–24 (2014)

7. Morais, R., et al.: Sun, wind and water flow as energy supply for small stationary data acquisition platforms. Comput. Electron. Agric. **64**, 120–132 (2008)
8. Weddell, A., et al.: Modular plug-and-play power resources for energy-aware wireless sensor nodes. In: Proceedings of the SECON 2009 Conference, pp. 1–9 (2009)
9. Magno, M., et al.: Smart power unit with ultra low power radio trigger capabilities for wireless sensor networks. In: Proceedings of the DATE Conference, pp. 75–80 (2012)
10. Panda, P.R., Silpa, B.V.N., Shrivastava, A., Gummidipudi, K.: Power-Efficient System Design, pp. 11–39. Springer, New York (2010). Chap. 2
11. Bohr, M.: Silicon technology leadership for the mobility era. In: Intel developer forum (2012)
12. Dghais, W., Rodriguez, J.: UTTB FDSOI back-gate biasing for low power and high-speed chip design. In: Mumtaz, S., Rodriguez, J., Katz, M., Wang, C., Nascimento, A. (eds.) WICON 2014. LNICST, vol. 146, pp. 113–121. Springer, Heidelberg (2015). doi:10.1007/978-3-319-18802-7_16
13. Dghais, W., Rodriguez, J.: Empirical modelling of FDSOI CMOS inverter for signal/power integrity simulation. In: IEEE Proceedings of the 2015 Design, Automation and Test in Europe Conference, Grenoble, France (2015)
14. Bailey, A., Zahrani, A.A., Fu, G., Di, J., Smith, S.C.: Multi-threshold asynchronous circuit design for ultra-low power. J. Low Power Electron. **4**, 337–348 (2008)

AutoDrop: Automatic DDoS Detection and Its Mitigation with Combination of OpenFlow and sFlow

Faisal Shahzad[1,2], Muazzam A. Khan[1(✉)], Shoab A. Khan[1], Saad Rehman[1], and Monis Akhlaq[2]

[1] College of EME, National University of Science and Technology, Islamabad, Pakistan
ifaisalshahzad@gmail.com, {muazzamak,shoabak}@ce.ceme.edu.pk
[2] Deltasoft Technologies, Rawalpindi, Pakistan

Abstract. World is emerging into global village with the support of internet connectivity. With the help of this connectivity, it also made everyone subject of being compromised. Many organizations' confidential data and numerous online services become victim of cyber-attacks. Different researches and innovations have been made for making network secure but commercial routers limit them to deploy custom security algorithms in real network. Recently, researchers succeed to innovate a novel protocol OpenFlow in Software Defined Networks. Taking advantage of this innovation we utilized OpenFlow to analyze real-time traffic, detect DDoS attack and mitigate attack. In this paper, we proposed a methodology to automatically detect different type of DDoS attacks within few seconds of occurrence using sampling techniques for continuous monitoring site-wide traffic and block attacking source with the help of OpenFlow protocol.

Keywords: DDoS · SDN · OpenFlow · sFlow · Security

1 Introduction

Many network applications, now-a-days are real time in nature. In a real time application, a good response time (efficient) and high latency rate of users requests are expected. Meeting the users expectations and needs of efficient response time is a major issue to be addressed in real time networks, where the traffic rate is also high, but it becomes really difficult to maintain a reasonable traffic flow in such applications when the unwanted software attacks are thrown to the system. These unwanted attacks make a network system slow or sometimes totally unavailable for its intended users. Our research objective is to address this Distributed Denial of Service (DDoS) problem in Software Defined Network (SDN) architecture.

Software Defined Network (SDN) has emerged in last two decades since 1990. Researchers were eager to make network programmable. In start, from 1990 to

© ICST Institute for Computer Sciences, Social Informatics and Telecommunications Engineering 2017
J. Ferreira and M. Alam (Eds.): Future 5V 2016, LNICST 185, pp. 112–122, 2017.
DOI: 10.1007/978-3-319-51207-5_11

2000, many small functions/scripts were developed to automate the network. From 2000 to 2007, researchers focused on separating control plane and data plane. In 2006, Martin Casado, PhD student at Stanford University in Silicon Valley developed something called Ethan. Later on, Stanford and University of California, Berkeley did a joint research and standardize protocol with the name of OpenFlow [1].

OpenFlow is an open standard that enables researchers to run experimental protocols in the campus networks. OpenFlow is used at commercial level and implemented by different vendors. It is embedded in different Ethernet switches, routers and wireless access points. It provides a research platform to run experiments without exposing the networks internal details. OpenFlow facilitates the researchers to innovate routing and switching protocols in networks. OpenFlow commercially being utilized in many applications like virtual machine mobility, high security networks and next generation IP based mobile networks [2].

2   Literature Review

Yao et al., proposed Virtual Source Address Validation Edge (VAVE), a method for securing network from spoofed IPs. They propose the use of OpenFlow devices rather than SAVI devices [3]. They pointed out that SAVI devices are good to detect spoofed IPs but each SAVI device is working independently without collaborating or using other device knowledge because they cannot communicate with each other, eventually which cause recalculation on each device again and again. In their solution, they were used OpenFlow devices with central controller and form a perimeter, when packets enter to perimeter, first OpenFlow device apply validation filters on packet using NOX controller, remaining all just used that information. Shin et al., proposed CloudWatcher, another network security solution that differ from VAVE. They proposed a simple scripting language to help network operators in defining policies. In this design, OpenFlow controller does not have any security module; instead they used other appliance for network security such as intrusion detection system. In this system, OpenFlow captures incoming network packets and forwards them to security appliances that inspect all of them [3]. Instead of OpenFlow controller, Kumar et al., proposed an OpenFlow switch for intrusion detection by adding two more tables for attacker IP (IDS IP) and signatures of malicious attacks. First they look into IDS IP table, if they found packet IP there then they simply drop the packet. If match is not found they look into IDS signatures table for checking its packet malicious or not. If packet found malicious they add it to IDS IP table for future use [4]. This solution greatly helps in: (1) Real-time packet inspection and validation, (2) Real-time attack mitigation, (3) Secure and intelligent network. On the other hand, this solution also impacts on: (1) Number of packets processing per second, (2) Traffic flow per second, (3) Cause bottle-neck in the network, (4) Significant impact on network performance.

Li et al. proposed DrawBridge, based on the assumption that if customer's controller can communicate with ISP controller. This enables customers to subscribe for ISP's traffic engineering service. Customer express its traffic engineering policies to DrawBridge controller in ISP, the DrawBridge controller further pushes these policies to SDN switches deployed in ISP network to filter traffic or another DrawBridge controller in the ISP upstream [6]. They preferred to throttle traffic rather than block attacking node and also suggest to install rules on ISP hosted OpenFlow controller. They are throttling traffic using Open-Flow protocol instead of dropping useless traffic. By this way useless traffic will be still coming in network and will be making resources busy in useless work. Based on same assumptions as DrawBridge, Sahay et al. proposed autonomic DDoS mitigation using software defined network. They proposed that DDoS mitigation can be autonomous application, as a service which can be consumed by customer network and multiple ISP networks. Both customer and ISP should have their own DDoS detection engine, can alert each other about attacks and use middle box mitigation application [7]. They proposed a novel change in the framework; this can greatly reduce the effort of developing mitigation application individually. But there will be chances of one-point failure.

Mousavi, researched on attack detection using controller rather than network because OpenFlow controller is back bone of SDN architecture. If someone targets OpenFlow controller by spoofed IP packets, then controller resources will be consumed by spoofed IP, it may cause out of service and by this way SDN architecture can be collapsed easily. He proposed entropy based light weight solution in the controller by just adding two code functions. But he does not focus on network. If someone attack all hosts in the network then this system will not be able to detect attack [8]. Mehdi et al., proposed traffic anomaly detection on SDN environment. They used multiple anomaly detection algorithms for validating their suitability in small office/home office environment. Author suggests that the decentralized control of distributed low-end network devices using OpenFlow, can efficiently detect network anomalies and limit network security problems. They left mitigation of detected network anomalies on their future work [9].

Braga et al., proposed a light weight DDoS flooding attack detection using NOX/ OpenFlow. They used Self Organizing Map (artificial intelligence) algorithm for dynamically identifying DDoS attack [10]. Proposals for secure SDN are not limited. Shin et al., focused on solution for developing and deploying complex OpenFlow security applications in much easier and rapid way [11]. Phaal et al., demonstrates commercially available network monitoring tool sFlow; which is helpful for monitoring traffic in data network containing switches and routers. sFlow uses sampling techniques for continuous monitoring site-wide traffic for high speed switched and routed network [12]. Rehman et al., has proposed a flow monitoring tool OF@TIEN for network wide traffic visibility using sFlow monitoring tool. SDN flow monitoring application gets slice flow definitions from OpenFlow controller, loads them into sFlow-RT, fetches summary statistics and feeds them to Graphite real-time charting tool. Our monitoring system also enables us to monitor GRE tunnels which are used to isolate traffic of tenant networks [13].

Looking at all above identified limitations, we have experimented an architecture with a combination of OpenFlow Northbound API, Kinetic (OpenFlow) Controller and sFlow (efficient network monitoring tool) with much higher network traffic up to 130,000 packets per seconds.

3   Proposed Solution

We propose real time traffic monitoring mechanism entering to customer network from ISP provider or internet exchange (IX). It monitors the statistics of traffic, passing through OpenFlow enabled switches, checks for anomalies and only forwards those packets which are from authentic users. If any malicious user found, it blocks the source and make it inaccessible for all.

Our proposed solution for DDoS attacks detection and their mitigation is based on: (1) Separation of network monitoring, attack detection and attack mitigation, (2) Compatibility with any OpenFlow-enabled switches, (3) Efficient network monitoring for attack detection, (4) Real-time attack detection, (5) Real-time attack mitigation, (6) Scalability with varying traffic.

4   Architecture Overview

Our system comprises of three major components as shown in Fig. 1.

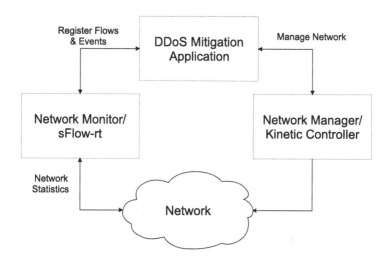

Fig. 1. DDOS mitigation solution using sflow-rt and OpenFlow.

4.1 Network Monitor

This module is responsible for real-time network monitoring and managing statistics which are prerequisite of DDoS detection. We used sFlow-RT as traffic statistics collector and analytics engine. It receives a continuous stream of data grams from network devices and converts them into actionable metrics that are accessible through REST APIs. REST APIs makes it easy for each application to configure flows, retrieve metrics, set thresholds, and receive notifications.

We have utilized sFlow-rt REST API for implementing sampling technique to continuous monitor site-wide traffic for high speed switched and routed network. DDoS mitigation application registers flows and specify threshold for generating alerts from sFlow-rt analytics engine using following curl commands described in Sect. 4.3.

4.2 Network State Manager

This module is responsible for managing traffic flows over the network and mitigating the identified attack. We have chosen Kinetic (OpenFlow) Controller due to its concise, intuitive way of expressing dynamic network policies. This is an event driven OpenFlow controller that allows to dynamically changing network behavior based on various types of network events. Kinetic controller manages network state based on Finite State Machine (FSM) mechanism, which gives a concise logical understanding for making the policies [14]. We have specified two network states and two policies i.e. Infected, Not Infected, identity and drop respectively. Infected or not infected specifies, whether source of coming request is infected or not and should we treat this packet as normal or simply drop because it's from infected source as shown in Fig. 2.

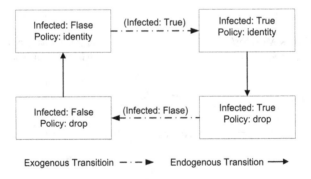

Fig. 2. Finite state machine (FSM) of network.

4.3 DDoS Mitigation Application

This module has a key responsibility for real-time DDoS detection and its real-time mitigation. This module communicates with both sFlow and Kinetic Controller using their REST APIs. It registers flows and threshold in sFlow-rt using its REST API. Multiple flows can be registered for different types of attacks with their corresponding thresholds. sFlow-rt generates events if traffic meets specified threshold. This module update network by passing Kinetic Controller about host and their status. Kinetic Controller drops all the packets coming from that specific host. Our system is implemented in python with the combination of node.js using JavaScript. Following steps demonstrates REST commands used to monitor and control the network:-

Define Flows

```
Curl -H "Content-Type:application/json" -X PUT --data "{keys:'
ipsource,ipdestination', value:'frames', filter:'sourcegroup=
external&destinationgroup=internal'}"http://localhost:8008/
flow/incoming/json
```

Define Thresholds

```
curl -H "Content-Type:application/json" -X PUT --data "{metric:'
incoming',value:1000}"http://localhost:8008/threshold/incoming/json
```

Receive Threshold Event

```
[{"agent":"10.0.0.50","dataSource":"4","eventID":5,"metric":"
incoming","threshold":1000,"thresholdID":"incoming","timestamp
":1357169369479,"value": 1531.149418835524 }]
```

Monitor Flow

```
[{"agent":"10.0.0.50", "dataSource":"4", "metricName":
"incoming", "metricValue":1582.93965044338071, "topKeys":
[{"key": "192.168.1.1, 10.0.0.50","updateTime":1357169662500,
"value":1582.93965044338071}, {"key": "192.168.1.4,10.0.0.50",
"updateTime":1357169665500,"value": 46.552918457198984 } ],
"updateTime": 1357169665500 }]
```

Deploy Control

```
../pyretic/pyretic/kinetic/json_sender.py -n infected 1 infected --flow='
{srcip=10.0.0.50}' -a 127.0.0.1 -p 50001
```

5   Results

In this experiment we are using Ping Flood attack to elaborate execution of our system. First, we had built a virtual network topology; having one switch with three hosts (i.e. h1, h2, h3). Then we flood h2 with Ping Flood (100,000 packets per second), sFlow-RT Fig. 3 shows that ping flood attack generates around 80,000 packets per second traffic rate.

Then we started our mitigation application in mininet and again generate Ping Flood attack, sFlow-RT quickly detected ping flood attack and notified the mitigation application. The mitigation application cross verified attack with specified threshold and sent notification to Kinetic controller.

Kinetic controller pushes rule to Open vSwitch using OpenFlow which instantly starts dropping packets. Figure 4 shows, our system quickly detected when traffic exceeds specified threshold and immediately mitigated attack in the tenth part of second rather than reaching a peak of 80,000 packets per second. Attack is limited to a peak of 550 packets per second. We choose 500 packets

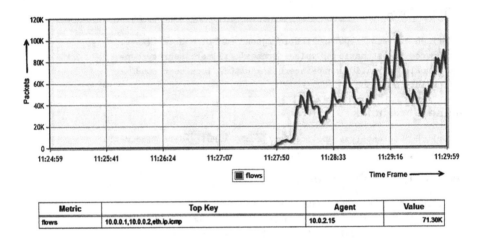

Metric	Top Key	Agent	Value
flows	10.0.0.1,10.0.0.2,eth.ip.icmp	10.0.2.15	71.30K

Fig. 3. Network on attack without presence of our application.

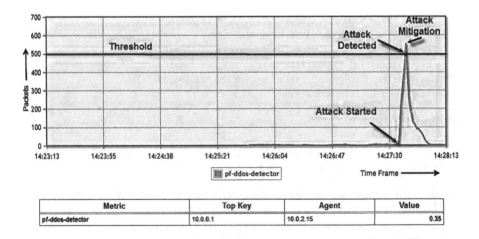

Metric	Top Key	Agent	Value
pf-ddos-detector	10.0.0.1	10.0.2.15	0.35

Fig. 4. Attack detection and mitigation in presence of our application.

per second threshold for demonstration purpose. This can be changed as per network traffic flow.

Attack detection time is inversely proportional to attack flow size and directly proportional to threshold. Attack mitigation is independent of threshold and attack size. It is responsible for blocking attacking source within a second. In addition, we have applied multiple control functions in parallel based, on the sFlow data feed, to detect multiple types of DDoS attacks such as Ping Flood, SYN and Ping of Death. The Table 1 summarizes detection and mitigation time vs. flow size of our experiment:

Table 1. Flow detection and mitigation time

Flow size(packets per second)	Threshold	Detection and mitigation time (s)
10	100	100–120
100	100	85–90
150	100	23–25
200	100	18–20
1000	1000	1.819–2.017
10000	1000	0.753–1.149

The detection times shown in Table 1 with different sampling values are shown in Table 2.

Table 2. sFlow sampling statistics

Link speed	Large flow	Sampling rate	Polling interval
10 Mbit/s	≥ 1 Mbit/s	1-in-10	20 s
100 Mbit/s	≥ 10 Mbit/s	1-in-100	20 s
1 Gbit/s	≥ 100 Mbit/s	1-in-1,000	20 s
10 Gbit/s	≥ 1 Gbit/s	1-in-10,000	20 s
40 Gbit/s	≥ 4 Gbit/s	1-in-40,000	20 s
100 Gbit/s	≥ 10 Gbit/s	1-in-100,000	20 s

6   Discussion and Comparison

There are various kinds of DDoS attacks; Ping Flood, Ping of Death and SYN Flood are most famous attacks in recent history. Many researchers have worked on different ways to identify DDoS attacks but most of them keep their focus on just attack detection rather than mitigating the attack source as well. Mitigation of attack source is also as important as its identification. Mostly they have focused on the attack detection without Worrying about the performance of the network [3–6].

Table 3. Comparison with literature work

	Impact on performance	Ping flood	Ping of death	SYN flood	Spoofed IPs	Attack mitigation
VAVE [3]	-	✗	✗	✗	✓	✗
CloudWatcher [4]	✓	-	-	-	-	✗
Kumar et. al extended OpenFlow switch [5]	✓	✓	✓	✗	✗	✓
DrawBridge [6]	-	-	-	-	✗	✓
Autonomic DDoS mitigation [7]	-	✓	✓	✓	-	✓
Light weight DDoS flooding attack detection [10]	✗	✓	✓	✓	✗	✗
FRESCO framework [11]	-	-	-	-	-	-
Mehdi et al. anomaly detection module [9]	✓	✓	✓	✗	✗	✗
Early detection of DDoS [8]	✓	✗	✗	✗	✓	✗
Proposed system	✗	✓	✓	✓	✗	✓

Braga et al., proposed a novel solution for identifying network anomalies using self-organizing map but didn't focus on attack mitigation [10]. FRESCO provided a full development framework for developing network security applications which can be easily deployed over OpenFlow enabled network [11]. Mehdi et al. also proposed a solution using different algorithms using OpenFlow. They monitor and process each packet for identifying whether it is malicious or not? This approach processes 600 packets per second [9]. If someone attacks whole network, then controller will not be able to detect the attack effectively as shown in Table 3.

Our System monitors network asynchronously and gather all traffic statistics using sFlow-RT. Our system identifies three most famous DDoS attacks; Ping Flood, Ping of Death and SYN Flood on real time with performance varying from 80,000-130,000 packets per second.

7   Conclusion

In this paper, we have evaluated Software Defined Network (SDN) for mitigating a huge network threat by DDoS attacks using OpenFlow protocol. Our study demonstrated that entire network monitoring - on periodic basis including tenth of thousands of flows - does not scale for high traffic environment. Moreover, using this technique a small medium flood attack may cause denial of service. We proposed a solution which: (1) reduces data gathering overhead by using sampling technique implemented through sFlow protocol, (2) detects anomalies using sFlow-rt analytics engine events, handled in most efficient JavaScript language (3) mitigates anomalies using OpenFlow protocol. We have offloaded OpenFlow for network monitoring with sFlow-rt, it have/leaves impact on network traffic speed. Our system performance is not only comparable with that of OpenFlow technique for low traffic rate but also reliable for high traffic networks as well. Our Proposed and implemented system handles real time traffic more efficiently than the prevailing techniques.

References

1. Feamster, N., Rexford, J., Zegura, E.: The road to SDN: an intellectual history of programmable networks. ACM SIGCOMM Comput. Commun. Rev. **44**(2), 87–98 (2014)
2. McKeown, N., Anderson, T., Balakrishnan, H., Parulkar, G., Peterson, L., Rexford, J., Shenker, S., Turner, J.: OpenFlow: enabling innovation in campus networks. ACM SIGCOMM Comput. Commun. Rev. **38**(2), 42–47 (2008)
3. Yao, G., Bi, J., Xiao, P.: Source address validation solution with OpenFlow/NOX architecture. In: Proceedings of International Conference on Network Protocol (ICNP), pp. 7–12 (2011)
4. Shin, S., Gu, G.: CloudWatcher: network security monitoring using openflow in dynamic cloud networks. In: Proceedings of International Conference on Network Protocol (ICNP), pp. 1–6 (2012)
5. Kumar, S., Kumar, T., Singh, G., Nehra, M.S.: Open flow switch with intrusion detection system. Int. J. Sci. Res. Eng. Technol. (IJSRET) **1**, 1–4 (2012)
6. Li, J., Berg, S., Zhang, M., Reiher, P., Wei, T.: DrawBridge-software-defined DDoS-resistant traffic engineering. In: SIGCOMM 2014, pp. 591–592 (2014)
7. Sahay, R., Blanc, G., Zhang, Z., Debar, H.: Towards autonomic DDoS mitigation using software defined networking (2015, to appear)
8. Mousavi, S.M.: Early detection of DDoS attacks in software defined networks controller. Thesis, Carleton University, Ottawa, Ontario (2014)
9. Mehdi, S.A., Khalid, J., Khayam, S.A.: Revisiting traffic anomaly detection using software defined networking. In: Sommer, R., Balzarotti, D., Maier, G. (eds.) RAID 2011. LNCS, vol. 6961, pp. 161–180. Springer, Heidelberg (2011). doi:10.1007/978-3-642-23644-0_9
10. Braga, R., Edjard, M., Passito, A.: Lightweight DDoS flooding attack detection using NOX, OpenFlow. In: LCN 10 Proceedings of the IEEE 35th Conference on Local Computer, pp. 408–415 (2010)
11. Shin, S., Porras, P., Yegneswaran, V., Fong, M. GU, G., Tyson, M.: FRESCO: modular composable security services for software-defined networks. In: Proceedings of Network and Distributed Security Symposium (2013)
12. Phaal, P., Panchen, S., McKee, N.: InMon corporations sFlow: a method for monitoring traffic in switched and routed networks. IETF, RFC 3176, pp. 1–31 (2001)

13. Ur Rehman, S., Song, W.-C., Kang, M.: Network-wide traffic visibility in OF@TEIN SDN testbed using sFlow. In: Network Operations and Management Symposium (APNOMS), pp. 1–6. IEEE (2014)
14. Kim, H., Reich, J., Gupta, A., Shahbaz, M., Feamster, N., Clark, R.: Kinetic: verifiable dynamic network control. In: USENIX NSDI (2015)

The Role of Vehicular Cloud Computing in Road Traffic Management: A Survey

Iftikhar Ahmad$^{(\boxtimes)}$, Rafidah MD Noor, Ihsan Ali,
and Muhammad Ahsan Qureshi

Faculty of Computer Science and Information Technology,
University of Malaya, Kuala Lumpur, Malaysia
Ify_ia@yahoo.com

Abstract. The vehicular cloud computing (VCC) is an emerging technology that changed the vehicular communication and underlying traffic management applications. The underutilized resources of vehicles can be shared with other vehicles over the VANET to manage the road traffic more efficiently. The cloud computing and its capability of integrating and sharing resources, plays potential role in the development of traffic management systems (TMSs). This paper reviews the VCC based traffic management solutions to analyze the role of VCC in road traffic management. Particularly, an analysis of VANET based and VCC based TMSs is presented. To explore, the VANET infrastructure and services, a comparison of VCC based TMSs is provided. A taxonomy of vehicular clouds is presented, in order to identify and differentiate the type of vehicular cloud's integration. Potential future challenges and their solutions in respect of emerging technologies are also discussed. The VCC is envision to play an important role in further development of intelligence transportation system.

Keywords: VANET · Vehicular cloud computing · Traffic management systems · Traffic flow control

1 Introduction

The Vehicular ad hoc network (VANET) [1, 2] is a subset of MANET. But VANET behaves differently because VANET has different properties like known routes, high speed and mobility [1–5]. VANET enables vehicles to communicate and share data wirelessly. The increasing number of vehicles on the road increasing the traffic load on transportation infrastructure, thus, making the driving unsafe and uncomfortable. The existing transportation system need to be modified, in order to make traffic safe [3] and efficient. It means new traffic management solutions are required to handle the challenge.

Usually, when congestion happens, the road infrastructure seems too small to handle the larger demand. Nowadays, the vehicle manufacturers wish to provide all capabilities to improve road safety and infotainment services. In order to provide these services, intelligent transportation systems (ITS) [4–6] support different traffic applications. These applications includes traffic safety, non-safety and flow efficiency applications. VANET support these traffic applications and it is main part of ITS.

© ICST Institute for Computer Sciences, Social Informatics and Telecommunications Engineering 2017
J. Ferreira and M. Alam (Eds.): Future 5V 2016, LNICST 185, pp. 123–131, 2017.
DOI: 10.1007/978-3-319-51207-5_12

The vehicle manufacturers, research communities and government authorities are making their efforts toward creating a standardized platform for vehicular communications. Consortium are formed, by multinational companies to increase efforts to tackle the problem of traffic management.

It is worth mentioning that recent improvements in software, hardware and communication technologies empower the design and development of cloud computing technology. The capability of cloud computing like dynamic resource integration and sharing, plays an important role in the development of emerging TMSs. The improved data and resource sharing among vehicles leads to VCC. VCC provides access to the dynamically configurable and integrated vehicular underutilized resources. This make the numbers of new applications based on VCC to be developed in order to provide various services to the drivers and passengers [6]. The VCC gives birth to a new pool of services which is a paradigm shift in the development of vehicular TMSs. This influences the development of new TMS potentially. An analysis of the role of VCC is provided in Sect. 3.

Now, various resources of vehicles are available to the end users [7]. The traffic data processed over data centers or over remote server, to provide information back to vehicles and passengers to control the movement of vehicles. Communications in VANET are generally classified into vehicle-to-vehicle (V2V) and vehicle-to-Infrastructure (V2I) communication. Vehicle communicate with other vehicles and Road Side Units (RSU). RSUs are intelligent devices, process data and send information to other RSUs as well. Our contributions in this paper includes;

- Reviewing emerging VANET traffic management applications
- Services and infrastructure based analysis of VCC based traffic applications
- A basic taxonomy of vehicular clouds
- Potential future challenges and issues

The rest of the paper is organized as follows. The review of VANET based TMSs and comparative analysis are described in Sect. 2. The analysis of VCC based traffic management systems, vehicular cloud's taxonomy and comparison are discussed in Sect. 3. Section 4 gives the future challenges of traffic management and finally the conclusion is drawn in Sect. 5.

2   Traffic Management Based on VANET

The VANET is the fundamental part of ITS which transforms the way of driving on the road. Today, driving on the road is more secure, safe and comfortable. Many efforts are made to reach these objectives, however, VANET's drawbacks such as high mobility of the vehicle and security issue does not allow researchers to meet these objectives [36]. The mobile internet and social networking in vehicles brings people and drivers more close to each other. Today, cars and vehicles are furnished with communication, computing and sensing devices, and universal networks such as internet. The driving experience is more enjoyable, comfortable, safe and environmental friendly than past but there are a lot of new paradigms to explore. The on board computing abilities, the vehicles are furnished with, are not fully utilized by the applications mentioned above.

The TMSs are broadly divided into three categories V2V (infrastructure-less), V2I (infrastructure-based) and Hybrid. Further, on the basis of control strategy TMSs are categorized as adaptive and predictive. Table 1 shows the classification of recent TMSs.

Table 1. Classification of TMSs

VANET infrastructure	Traffic flow control strategy	
	Adaptive	Predictive
	Use a control strategy that adapts changes based on actual traffic demands	Use data analytics to determine potential future locations of congestion and traffic flow
V2V	[8–10]	[11]
V2I	[12, 13]	[14–16]
Hybrid	[17–19]	[20–22]

TMSs make use of few common schemes like traffic estimation (density), historic traffic information, future predictions, and more recent schemes like platooning (grouping). The traditional traffic lights uses static time limits at the traffic signal intersections. The dynamically traffic lights can be developed based on the number of vehicles in a lane and by giving priority to emergency vehicle to surpass the signal quickly [23].

In order to cater broadcasting storm, traffic information has to be limited in detail [24]. This can be done by exploiting V2V communication, by apply data aggregation techniques to limit bandwidth use and maintain scalability. To avoid the overlapping of aggregates for the same area, there must be a scheme. A better aggregation scheme exploit V2V communication more effectively. One of the solutions is formation of groups of connected vehicles (clusters) [25]. The formation of cluster should be done intelligently that avoids collision among clusters. The most appropriate V2V scheme must calculates and detects the level of congestion in distributed way without the support of any infrastructure and additional information [8]. The adaptive broadcasting scheme should be utilized which gives awareness to at micro level. The congestion quantification is good for low VANET penetration rate.

ITS provide several features like Multi-input and multi-output capability which can be utilized to form multi-objective function. Such, multi-objective function take multiple parameters to provide congestion control, vehicle re-routing and planning. As the system uses the feedback information measured from the real-time vehicular traffic, it works in a closed loop manner. Factors such as vehicle velocity, vehicle position and distance between vehicles as well as between vehicles and RSUs greatly affect the performance of TMSs. These factors also affect the reliability and delay of links.

3   Vehicular Cloud Computing in TMSs

An important property of cloud computing is that no investment is needed because instead of buying, resources and services are rented on demand. Vehicular cloud computing services of a particular cloud are dependent on the purpose for which the cloud is formed [26, 27]. Generally, vehicular clouds provide services [28] such as platform as a Service (PaaS), Infrastructure as a Service (IaaS), Software as a Service (SaaS), Application as a Service (AaaS), and storage as a service (STaaS). The explanation of these services is already there in literature like [26–29].

Vehicles can subscribe cloud provided services on demand. By connecting OBUs through wireless networks such as 3G/4G-LTE and Wi-Fi networks, users can obtain almost unlimited computing power and storage from the cloud. The VCC improves the collection, processing and dissemination of traffic related data. VCC integrates and coordinates the available vehicular resources and enables the road traffic management in better way. Which reduces risk of life, cost and time.

3.1   Taxonomy of Vehicular Clouds

The vehicular clouds have different types. On the basis of purpose for which a TMS form a cloud, the vehicular clouds are classified into two main classes named as V2V clouds and V2I clouds as shown in Fig. 1. The classification is done on the basis of services for which cloud is to be formed, infrastructure used and involvement of third party clouds (Internet and other commercial clouds).

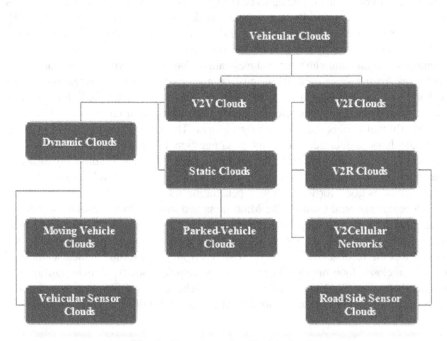

Fig. 1. Taxonomy of vehicular clouds

V2V Clouds. V2V clouds are formed by exploiting V2V type of communication infrastructure (DSRC). Dynamic clouds are formed by vehicles on the roads or in the parking lot, for typical service needed by underlying TMS. Like vehicles on the road form cloud to know the status of each other and to make intelligent decision regarding rout planning. Same as the vehicles in the parking lot provide storage and processing services named as static cloud.

These type of clouds rely on V2V communication infrastructure. Vehicular sensor clouds are formed for the purpose of vehicle or road traffic monitoring. The sensors within the vehicle gets vehicle related data, other vehicles in the cloud can take these services. For example a vehicle for behind the intersection can send query to the camera of a vehicle which is nearer to the intersection to get a real time picture of intersection. These type of services can be given through vehicular sensor clouds by utilizing sensors those are associated with vehicles.

V2I Clouds. In this type of vehicular clouds, the vehicles make use of road side infrastructural communication networks. These communication networks include DSRC, WIFI, 3G/LTE. If vehicles in cloud rely on RSUs for control information then the cloud called as V2R clouds and if vehicles in cloud rely on 3G/LTE networks then the cloud called as V2Cellular cloud. Both of these cloud types have their own pros and cons and it depends on the purpose for which clouds are being formed. As for larger geographical area the V2Celluar clouds are useful and for smaller coverage area V2R clouds are better. RSUs, intersections and special entry points are often equipped with traffic monitoring sensors, radars and cameras.

These sensor can work together to form a cloud to share real time information with TIS and vehicles. These type of clouds are known as road side vehicular sensor clouds. The services like participatory sensing and cooperative sensing are the main purpose of these clouds.

The identified types of vehicular clouds collaborate and integrates with each other in order to provide different services to users. Cloud cooperation and collaboration is emerging concept which may lead to new type of services.

3.2 Comparative Study of VCC Based TMSs

A comparison of some VCC based TMSs is presented in Table 2. Comparison is made on the basis of whether the TMSs is using V2V or V2I infrastructure, mitigating traffic congestion or providing flow control and type of services offered.

In "PaaS "the vehicular clouds provides a platform for other related services like content downloading and sharing. This also provides platform for traffic management authorities to have access to all of the vehicles on the road. Other service like "StaaS" VCC enable us to have access to huge storage capacity which is distributed over large number of vehicles. For "CaaS" the VCC utilize processing power of vehicles and distribute data processing among number of vehicles participating in the cloud.

As explained in cloud taxonomy section multiple clouds cooperate with each other in order to share services among them. For example a V2V cloud collaborate with Internet cloud in order to have access to internet based remote server (traffic information system). These services are analyzed and a comparison is provided in Table 2. It is clear from the Table 2 that the trend is towards more internet cloud independency and not all the TMSs are really mitigating traffic congestion but just claiming.

Table 2. Comparison of VCC based TMS's proposals and prototypes

TMS article	V2I	V2V	Traffic flow control	PaaS	STaaS	CaaS	CSaaS
[29]	yes	yes	yes	yes	yes	yes	yes
[30]	yes	yes	yes	yes	yes	yes	no
[31]	yes		no	yes	yes	yes	no
[32]	yes		yes	yes	yes	yes	no
[33]	yes	yes	yes	yes	no	yes	yes
[34]	yes	yes	no	yes	no	yes	yes
[35]	yes	yes	yes	yes	yes	yes	no
[36]	yes	yes	yes	yes	yes	no	no
[37]	yes	yes	yes	yes	no	yes	yes
[38]	yes	yes	yes	yes	yes	yes	yes

TMSs are mostly relying on V2I infrastructure then V2V and on both infrastructures as well. All of the TMSs are claiming and providing PaaS and STaaS is not providing by all. The emerging services like cloud cooperative sensing as a service (CSaaS) are not common in recent known VCC based TMSs. Few of them is providing CSaaS only which is in the form of participatory type of sensing and it is not fully cooperative.

4 Traffic Management Challenges and Solutions

Few of the main challenges are discussed in following paragraphs. These challenges are not yet handled properly.

Resource Management: One of the important solutions to overcome these problems is to allocate the accurate amount of resources at right time as compared to the pre-allocation of resources. There is a need of the real time traffic related information. The VCC can help in finding the appropriate solution by using the available resource of vehicle without waiting for officials and DMCs.

Infrastructural Support in Evacuation: The VCC can provide infrastructural support during disaster and emergency situations. The disaster management authority can utilize VCC for evacuation. VCC provides the efficient information related to time, place and the availability of necessary resources such as food, water, shelter etc. The vehicles participating in the evacuation procedure forms vehicular cloud and coordinates with the rescue response teams.

Road Safety and Warning: If there is an incident/event occurs on the road then the vehicle inform/warn nearby vehicle and so on to the vehicular cloud regarding speed, location and direction of the incident/event. This early information or warning is very helpful for the vehicles so that they may decide/re-rout for the rest of the journey.

Intersection Congestion Management: Most of the time vehicles must have to, pass through the intersections. Because where there is a congested and highly populated area, there is a greater possibility of shopping malls, hospitals, schools and other frequently visiting buildings. The VCC can play a role in intersection management. The latest updated information on the vehicular cloud can provide an efficient solution to the drivers. This early information or warning is very helpful for the vehicles so that they may decide/re-route for the rest of the journey.

Communication Challenges: The communication among vehicles, vehicles to RSUs and vehicle to cloud (e.g. internet cloud) is critical. The vehicle's decisions related to safety and comfort depends on successful communication. The probability of successful communication depends on various factors like spectrum congestion, cost of internet use and on type of technology being used for communication.

Incorporation of IoVs: The impact of new technological advancements in the field of VANET, cloud computing and IoVs has not yet been fully realized. There is much need of the integration of such technologies to cope the challenges of the traffic congestion and transportation. Furthermore, how to balance the computations among local vehicles, vehicular clouds and Internet cloud so as to achieve different goals by IoVs. Future is of internet of things (IoTs) and big data, therefore the vehicular cloud cooperation is an intermediator in this paradigms shift. The devices and services collaboration among multiple clouds can be done by fully realizing the IoVs.

5 Conclusion

The VCC have great potential to change our lives on the road, by utilizing and sharing resources of vehicles with other vehicles to manage the traffic during congestion. The VCC provides an efficient enhancement to the message dissemination, traffic management and congestion control. The comparative analysis of VANET and VCC based TMSs is provided to show the purpose oriented scope of vehicular clouds and their use in road traffic management. The future challenges and their solution in terms of vehicular cloud cooperation is predicted. The VCC is envision to play an important role in further development of intelligence transportation system. The VCC and its emerging form of IoVs has great potential to derive the autonomous vehicles more efficiently.

References

1. Ahmad, I., Ashraf, U., Ghafoor, A.: A comparative QoS survey of mobile ad hoc network routing protocols. J. Chin. Inst. Eng. **39**, 585–592 (2016)
2. Ahmad, I., Rehman, M.U.: Efficient AODV routing based on traffic load and mobility of node in MANET. In: 2010 6th International Conference on Emerging Technologies (ICET), pp. 370–375. IEEE (2010)
3. Alam, M., Fernandes, B., Silva, L., Khan, A., Ferreira, J.: Implementation and analysis of traffic safety protocols based on ETSI Standard. In: 2015 IEEE Vehicular Networking Conference (VNC), pp. 143–150. IEEE (2015)
4. Fernandes, B., Alam, M., Gomes, V., Ferreira, J., Oliveira, A.: Automatic accident detection with multi-modal alert system implementation for ITS. Veh. Commun. **3**, 1–11 (2016)

5. Alam, M., Ferreira, J., Fonseca, J.: Introduction to intelligent transportation systems. In: Alam, M., Ferreira, J., Fonseca, J. (eds.). SSDC, vol. 52, pp. 1–17. Springer, Heidelberg (2016). doi:10.1007/978-3-319-28183-4_1
6. Shah, S.A.A., Ahmed, E., Xia, F., Karim, A., Shiraz, M., Noor, R.M.: Adaptive beaconing approaches for vehicular ad hoc networks: a survey. IEEE Syst. J. **PP**, 1–15 (2016)
7. Anisi, M.H., Abdullah, A.H.: Efficient data reporting in intelligent transportation systems. Netw. Spat. Econ. **16**(2), 623–642 (2016). doi:10.1007/s11067-015-9291-9
8. Milojevic, M., Rakocevic, V.: Distributed road traffic congestion quantification using cooperative VANETs. In: 2014 13th Annual Mediterranean Ad Hoc Networking Workshop (MED-HOC-NET), pp. 203–210. IEEE (2014)
9. Chen, W., Guha, R., Lee, J., Onishi, R., Vuyyuru, R.: A multi-antenna switched links based inter-vehicular network architecture. In: 2009 IEEE Vehicular Networking Conference (VNC), pp. 1–7. IEEE (2009)
10. Gupte, S., Younis, M.: Vehicular networking for intelligent and autonomous traffic management. In: 2012 IEEE International Conference on Communications (ICC), pp. 5306–5310. IEEE (2012)
11. Tak, S., Kim, S., Yeo, H.: A study on the traffic predictive cruise control strategy with downstream traffic information. IEEE Trans. Intell. Transp. Syst. **17**(7), 1932–1943 (2016). doi:10.1109/TITS.2016.2516253
12. Yugapriya, R., Dhivya, P., Dhivya, M., Kirubakaran, S.: Adaptive traffic management with VANET in V to I communication using greedy forwarding algorithm. In: 2014 International Conference on Information Communication and Embedded Systems (ICICES), pp. 1–6. IEEE (2014)
13. Djahel, S., Salehie, M., Tal, I., Jamshidi, P.: Adaptive traffic management for secure and efficient emergency services in smart cities. In: 2013 IEEE International Conference on Pervasive Computing and Communications Workshops (PERCOM Workshops), pp. 340–343. IEEE (2013)
14. Nafi, N.S., Khan, R.H., Khan, J.Y., Gregory, M.: A predictive road traffic management system based on vehicular ad-hoc network. In: 2014 Australasian Telecommunication Networks and Applications Conference (ATNAC), pp. 135–140. IEEE (2014)
15. Daniel, A., Paul, A., Rajkumar, N.: Embedded surveillance system for vehicular networks. In: 2015 2nd International Conference on Electronics and Communication Systems (ICECS), pp. 1635–1640. IEEE (2015)
16. Alrifaee, B., Granados Jodar, J., Abel, D.: Predictive cruise control for energy saving in REEV using V2I information. In: 2015 23rd Mediterranean Conference on Control and Automation (MED), pp. 82–87. IEEE (2015)
17. Lunge, A., Borkar, P.: A review on improving traffic flow using cooperative adaptive cruise control system. In: 2015 2nd International Conference on Electronics and Communication Systems (ICECS), pp. 1474–1479. IEEE (2015)
18. Xiao, L., Lo, H.K.: Adaptive vehicle navigation with en route stochastic traffic information. IEEE Trans. Intell. Transp. Syst. **15**, 1900–1912 (2014)
19. Wang, S., Djahel, S., McManis, J.: An adaptive and VANETs-based Next road re-routing system for unexpected urban traffic congestion avoidance. In: 2015 IEEE Vehicular Networking Conference (VNC), pp. 196–203. IEEE (2015)
20. HomChaudhuri, B., Vahidi, A., Pisu, P.: A fuel economic model predictive control strategy for a group of connected vehicles in urban roads. In: 2015 American Control Conference (ACC), pp. 2741–2746. IEEE (2015)
21. Liang, Z., Wakahara, Y.: City traffic prediction based on real-time traffic information for intelligent transport systems. In: 2013 13th International Conference on ITS Telecommunications (ITST), pp. 378–383. IEEE (2013)

22. Kamal, M., Taguchi, S., Yoshimura, T.: Efficient vehicle driving on multi-lane roads using model predictive control under a connected vehicle environment. In: 2015 IEEE Intelligent Vehicles Symposium (IV), pp. 736–741. IEEE (2015)
23. Collins, K., Muntean, G.-M.: Traffcon: An intelligent traffic control system for wireless vehicular networks. IET CIICT (2007)
24. Al-Sultan, S., Al-Doori, M.M., Al-Bayatti, A.H., Zedan, H.: A comprehensive survey on vehicular ad hoc network. J. Netw. Comput. Appl. **37**, 380–392 (2014)
25. Ramakrishnan, B., Nishanth, R.B., Joe, M.M., Selvi, M.: Cluster based emergency message broadcasting technique for vehicular ad hoc network. Wireless Netw., 1–16 (2015)
26. Gerla, M., Lee, E.-K., Pau, G., Lee, U.: Internet of vehicles: from intelligent grid to autonomous cars and vehicular clouds. In: 2014 IEEE World Forum on Internet of Things (WF-IoT), pp. 241–246. IEEE (2014)
27. Gerla, M.: Vehicular cloud computing. In: 2012 The 11th Annual Mediterranean Ad Hoc Networking Workshop (Med-Hoc-Net), pp. 152–155. IEEE (2012)
28. Whaiduzzaman, M., Sookhak, M., Gani, A., Buyya, R.: A survey on vehicular cloud computing. J. Netw. Comput. Appl. **40**, 325–344 (2014)
29. Wang, W.Q., Zhang, X., Zhang, J., Lim, H.B.: Smart traffic cloud: an infrastructure for traffic applications. In: 2012 IEEE 18th International Conference on Parallel and Distributed Systems (ICPADS), pp. 822–827. IEEE (2012)
30. Ma, M., Huang, Y., Chu, C.-H., Wang, P.: User-driven cloud transportation system for smart driving. In: 2012 IEEE 4th International Conference on Cloud Computing Technology and Science (CloudCom), pp. 658–665. IEEE (2012)
31. Arif, S., Olariu, S., Wang, J., Yan, G., Yang, W., Khalil, I.: Datacenter at the airport: reasoning about time-dependent parking lot occupancy. IEEE Trans. Parallel Distrib. Syst. **23**, 2067–2080 (2012)
32. Mershad, K., Artail, H.: CROWN: discovering and consuming services in vehicular clouds. In: 2013 Third International Conference on Communications and Information Technology (ICCIT), pp. 98–102. IEEE (2013)
33. Kumar, S., Shi, L., Ahmed, N., Gil, S., Katabi, D., Rus, D.: Carspeak: a content-centric network for autonomous driving. ACM SIGCOMM Comput. Commun. Rev. **42**, 259–270 (2012)
34. Gerla, M., Weng, J.-T., Pau, G.: Pics-on-wheels: photo surveillance in the vehicular cloud. In: 2013 International Conference on Computing, Networking and Communications (ICNC), pp. 1123–1127. IEEE (2013)
35. Abid, H., Phuong, L.T.T., Wang, J., Lee, S., Qaisar, S.: V-Cloud: vehicular cyber-physical systems and cloud computing. In: Proceedings of the 4th International Symposium on Applied Sciences in Biomedical and Communication Technologies, p. 165. ACM (2011)
36. Alazawi, Z., Alani, O., Abdljabar, M.B., Altowaijri, S., Mehmood, R.: A smart disaster management system for future cities. In: Proceedings of the 2014 ACM International Workshop on Wireless and Mobile Technologies for Smart Cities, pp. 1–10. ACM (2014)
37. Munst, W., Dannheim, C., Mader, M., Gay, N., Malnar, B., Al-Mamun, M., Icking, C.: Virtual traffic lights: Managing intersections in the cloud. In: 2015 7th International Workshop on Reliable Networks Design and Modeling (RNDM), pp. 329–334. IEEE (2015)
38. Khalid, O., Khan, M.U.S., Huang, Y., Khan, S.U., Zomaya, A.: EvacSys: a cloud-based service for emergency evacuation. IEEE Cloud Comput. **3**, 9 (2016)

Workshop on Internet of Things (IoT) meets Big Data and Cloud Computing

Architecture Proposal for MCloud IoT

Zeinab Y.A. Mohamed[1], Dalia M. Elsir[1], Majda O. Elbasheer[1],
and Jaime Lloret[2(✉)]

[1] Sudan University of Science and Technology, Khartoum, Sudan
zeinab_yassin@yahoo.com, dalia.m.elsir@hotmail.com,
majda.omer@gmail.com
[2] Universidad Politecnica de Valencia, Valencia, Spain
jlloret@dcom.upv.es

Abstract. The world is now heading towards an era shaped by things that able to act and interact through Internet. Despite of many IoT devices suffer from limitations regarding to storage, processing capability, and communication, IoT plays a major role providing a new set of applications and services. Cloud computing provides a supplement solution for IoT limitation. Integration of IoT and Cloud Computing is considered a new direction that both scientist and business are seeking for bringing new applications and benefits to existing applications and services. Moreover, the fast development of mobile devices produces powerful devices that are able to play many roles, creating better IoT scenarios. In this paper; we propose a new MCloud IoT architecture that works on an IoT environment, which is composed by mobile devices such as smart phones, tablets, and smart sensors. MCloud IoT architecture is designed to deliver the applications and services demanded by end users. Moreover, we have included a layered communication model for devices' communication. In our design we have taken into account the system performance and to provide QoS. We also analyze the benefits of our design. This new architecture provides a revolutionary vision that meets the future expectations of cloud systems.

Keywords: IoT · Cloud computing · Cloud of things · Cloud IoT architecture · Mobile devices

1 Introduction

The world is heading towards a new era (anything, anytime, anywhere) shaped by the Internet and things that have the ability to act and react through data [1]. This leads to massive transformation of almost every aspect in people's lives; the ways they act, learn, communicate, create new things, etc. Moreover, this evolution produces new technologies and paradigms which open even more opportunities and chances to pave the way to the shifting towards the smart things and services.

Internet of Things (IoT) paradigm is one of the key building block of this era, in this paradigm real objects called 'things' are smart enough to connect each other and to other systems and then to Internet. IoT provide many applications and services in addition of enabling ubiquitous computing. But things have some limitations in matter of storage and process capacity in complex computation [2].

© ICST Institute for Computer Sciences, Social Informatics and Telecommunications Engineering 2017
J. Ferreira and M. Alam (Eds.): Future 5V 2016, LNICST 185, pp. 135–145, 2017.
DOI: 10.1007/978-3-319-51207-5_13

Cloud computing is a technology that provides virtually infinite resources for data storage and processing. Cloud computing provides this through several services such as Infrastructure as a Service (IaaS), Platform as a Service (PaaS), and Software as a Service (SaaS) where end customers can choose easily the service to satisfy their needs [3]. Recently, there are appearing many task scheduling algorithms for cloud environments [4], which allow them to have high performance. Moreover, there are also appearing cloud systems to provide live video streaming [5].

IoT and Cloud computing are attracting a lot of attention despite of the differences between them. Many researchers believe that they are complementary and that may lead to new application scenarios in addition to improve the current ones. This line of researches toward the integration of IoT and Cloud computing have been referred as Cloud of Things.

Cloud of Things is a new paradigm which integrates IoT and Cloud Computing. It provides the features of ubiquitous IoT applications and services, and can be applied to smart homes, smart cities, health care, smart logistic, etc., in order to facilitate the end user life. It allows reaching anything anywhere at anytime, without taking care of where to store their data and how to perform any operation with no matter about the requirements of processing capacity or resources thanks to Cloud Computing. Cloud of Things should easily be able deliver IoT services, introduce specific performance, and QoS requirements to meet users demands.

The number of connected devices (especially smart mobile devices) is increasing continuously. In addition, they are massively spreading around the world. This number exceeds more than 10 billion and it is expected to reach 24 billion in 2020. That will lead to generate massive amount of data too. These data will require efficient ways to gather, store, process and extract knowledge from it. However, Cloud of Things using mobile phones is in its early stage, which means that a lot of work and research should be done to address different issues and challenges.

This paper aims to develop a new MCould IoT architecture; that works on IoT environment, which is composed of mobile devices such as smart phones, tablets, smart sensors, to deliver the application and services demanded by end users. In our design we have taken into account the system performance and to provide QoS.

This paper is structured as follows. Section 2 describes the related works. Section 3 presents the proposed MCloud IoT architecture. Section 4 describes the difference between our MCloud IoT proposal and a regular Cloud for IoT. We conclude this paper in Sect. 5.

2   Related Work

There are few works published about Cloud IoT. Some works address the integration of IoT and Cloud computing, but they still do not provide any detailed system or a standard solution to the challenges and issues of the area. Next, we describe the related works we have found.

In [6], authors present a Smart Gateway based communication plus Fog computing to offer smart communication with little computation overhead on core network. They handle real-time and delay sensitive applications by trimming and pre-processing the data before sending it to the cloud. Based on the tests performed of various performance parameters they believe that their proposed architecture will deliver a rich portfolio of services.

The authors of [7] proposed an architecture for Cloud of Things for sensing as a Service. The aim of the authors was to perform an in-network distributed processing system and an efficiently set up virtual sensor network on the top of a subset of the pre-selected IoT devices in order to provide a global platform for data analysis and decision making. Their proposed algorithms can realize virtual sensor networks with minimal physical resources, low complexity, and reduced communication overhead.

In [8], a conceptual platform and the defined key characteristics of Fog computing are presented. It is considered the appropriate platform for a number of critical IoT, where services and applications are handled at the edge of the network.

A Cloud provisioning model is proposed in [9]. It is an architecture designed to leverage from bridging Clouds with the IoT to meet user needs according to some guaranteed service levels. It also introduces things as infrastructure for Cloud like exploitation. Authors tried to address the ideas of the intersections between them where heterogeneous resources should be combined and abstracted according to tailored thing-like semantics paving the way for innovative and value-added services.

In [10], authors implement and test the behavior of a health monitoring system in the context of clouds and IoT. They introduced SimIoT toolkit with the utilization of short range and wireless communication devices to meet dynamic information processing where IoT devices schedule requests for services in private clouds.

The authors of [11] proposed an architecture model for medical information using IoT and cloud computing integration through the combination of technology monitoring and management information system of a hospital. Moreover, an effective algorithm is proposed for the medical monitoring application. The proposed remote monitoring cloud platform architecture model has been evaluated through an experimental analysis and simulation.

In [12], authors propose a smart gateway communication system for Cloud IoT architecture. The study aims to enhance service provisioning to the user and efficient utilization of resources using a smart gateway that performs several tasks such as data trimming and pre-processing before sending them to the cloud. In addition, they use Fog computing to alleviate the burden of the cloud. The paper also shows that normal communication can be made in real-time for delay sensitive applications.

The architecture proposed in [13] provides a simple, energy efficient, flexible, and secure scheme for a smart house based on Cloud of Things (CoT). The proposal ensures the security to transfer data through the proposed mechanism for smart housing. They consider different types of devices and their capabilities, the scalability of the smart house and the energy consumption.

In [14], a Model Driven Architecture (MDA) is used to develop Software as a Service (SaaS) to facilitate the mobile applications development by relieving developers from technical details.

The work shown in [15] aims to provide efficient access controls and sharing controls with slight virtualization overhead for a cloud of things architecture. Authors propose an Evolvable Cloud of things (ECO) middleware that makes use of a lease-based sharing control mechanism for enabling logical isolation and efficient sharing between multi-tenant applications through virtualization. The validation of the system performance confirmed that it reduces the effort and complexities when implementing and developing applications. The system also provides and effective sharing with a little virtualization overhead.

Along the reviewed related literature we have not seen any designed system that integrates mobile IoT devices and cloud computing.

3   MCloud IoT Architecture

Integration of IoT and Cloud computing is considered a new direction that both scientist and business seeking for and interest about to bring new applications and benefits to existing applications and services. Nowadays there are many powerful mobile devices (smart phones tablets, and even sensors) acting as things in Internet [16]. They gather data from the surrounding environment and store them locally or remotely for further processing. They exist in any environment, composing one of the best IoT scenarios. These devices can offer different services such as storage resource, a processing capability, a gateway to other network and/or Internet. On the other hand, the fact of having a cloud for the data and services of these devices include many constraints, which brings the need of research for providing Cloud IoT solutions based on the idea of those mobile devices to create a cloud.

The proposed MCloud of Things architecture allows mobile users to create their own cloud using a Cloud IoT application which implements the cloud agent in their devices. Then the users will be able to access the shared resources in the cloud such as storage, run some tasks, with certain specifications, virtually on any other mobile devices in IoT environment as shown in Fig. 1.

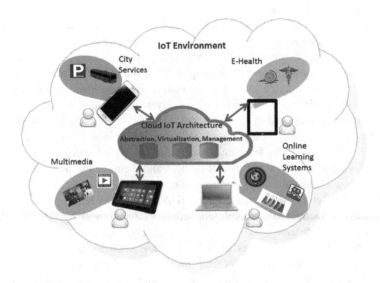

Fig. 1. Overview of MCloud IoT architecture.

3.1 Conceptual Layers View of MCloud IoT Architecture

The conceptual view of the proposed MCloud IoT architecture structure form 3 main layers as shown in Fig. 2:

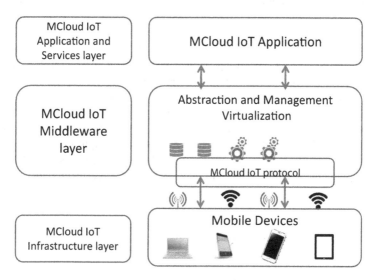

Fig. 2. Conceptual layers view of cloud IoT architecture.

- Infrastructure layer that handles the heterogeneity of the mobile devices;
- The middleware layer that is responsible of managing the resources of the cloud (virtualizes the resources, determines the roles of each device, extracts the resources of the tracking the status). It also contains the cloud agent which responsible of arranging the service such as resource identification resource discovery, transform the data into generic form to store it into the cloud and/or deliver it to other devices through the cloud. The application layer that contains the MCloud IoT application. It is the interface with the new cloud or other IoT application and services that will used by the devices participating into the MCloud IoT, such as resource allocation, perform processing, or deliver new services.

MCloud IoT architecture will help in managing IoT resources, allowing delivering new services to end user; for example providing the services from different devices and environment into the cloud will simplify the service delivery in IoT environment because in this case it will have an ubiquitous access for the users and it will extend the usage of the service into larger section of user.

The future rely on mobile devices as a key element to access, control, store, and mange different data through large set of applications that serve wide range of people needs. So, it will be a promising line to work more in the capabilities of these devices.

Users require more and more storage to store their data, more processing capacity to perform complex task. They want to be online everywhere at anytime. In order to achieve this expectation, new architectures and mechanisms are required.

A closer look into the IoT architecture layers is illustrated in Fig. 3. Things layer is the lowest layer and it represents different objects that perceive data from the surrounding environment such as mobile devices, sensors, objects with RFID tag,...

Connectivity layer: it is similar to network layer in OSI model. It includes a gateway to transfer the collected data into next layer through variety of wireless technologies and communication protocols such as Wifi, NFC, and Bluetooth. Thus, it has one interface connected to the things network and another to Internet.

Middleware layer: it provides an abstraction for the underlying infrastructure, dealing with different issue according to the heterogeneity, it responsible of service management such as service identification and discovery, tacking the status of the devices. It also handles the context management of the data.

Service layer: its purpose is to provide cloud services to the data such as storage of data, perform information processing and take decisions. In addition to that, it protects the data using suitable security mechanisms. It passes the output to the next layer.

Application layer: it presents the final form of data. It can process the data for large number of applications in different areas such as smart home, health care, etc.

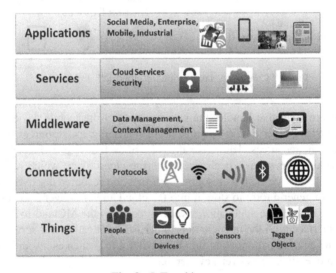

Fig. 3. IoT architecture

3.2   MCloud IoT Architecture Components and Functionalities

The new MCloud IoT architecture works on IoT scenario considering mobile devices as things of that environment. We call it Mobile Cloud IoT or MCloud IoT.

The components for new MCloud IoT architecture are shown in Fig. 4:

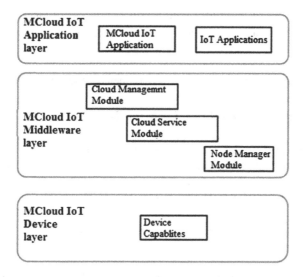

Fig. 4. MCloud IoT architecture components

- **The MCloud IoT application**: it is a mobile application installed into the mobile devices such as smart phones, tablets, laptops; it represents an interface to the MCloud IoT middleware.
- **The MCloud IoT middleware**: it is responsible of providing the abstraction and management through a number of modules describes as follow:
  - **The Node manager module**: responsible of determining the device role (master, resource provider, gateway), status (active or not), type of communication protocol.
  - **The cloud service module**: it is in charge of the identification of the resource such as resource discovery, resource allocation, storage resource, computing resource, communications resource, and other services defined by cloud users.
  - **The cloud management module**: it is responsible of monitoring the resources, cloud services and other components.
- **The MCloud IoT device layer**: it is responsible of determining the connected devices capacities such as CPU, RAM, and storage.

The proposed architecture is created in a form of wireless ad hoc network, thus, the users should install the MCloud IoT application to be able to implement the proposed architecture and then use it. The application implements the proposed protocol that allows the device to join the cloud and thus discover the services and resources provided on it, after that, it can choose the suitable service based on its needs.

The new MCloud IoT platform also allows the participation of devices to connect to other networks and transfer the data to/from them, even if one device doesn't have a direct gateway to that network. The MCloud IoT will assign one of the devices to act as a gateway for the request device. The new architecture will help in developing new application scenarios that benefit of the capabilities of mobile devices.

There are number of issues that may face the proposed architecture and it has to take into account developing it. These issues such as the heterogeneity of devices, protocol support, how this kind of cloud will interact with other existing clouds, resource allocation, resource identification, security and privacy, reliability, QoS provisioning will examine and cover in the development of the MCloud IoT architecture.

4 MCloud IoT Vs Regular Cloud for IoT

While a regular Cloud for IoT is a Cloud created by servers placed in Internet which store data from the things and provide services to them, a MCloud IoT is a cloud formed by the mobile devices acting as things in Internet. These concepts provide clear differences between their features and the environments where they can be more useful. In Table 1, we provide the main differences. We can observe that MClould IoT will benefit to those systems where the latency and jitter are critical.

Table 1. Comparison between MCloud IoT and regular cloud for IoT

	MCloud IoT	Regular cloud for IoT
Computing capacity	Regular	Very high
Energy/battery	Few	Very high
Storage	Regular	Very high
Bandwidth	Regular (the bottleneck is the "things" connection)	Regular (the bottleneck is the "things" connection)
Latency	Low	High
Jitter	Low	High

It is well known that the most critical issues in cloud computing are its high delay and jitter values [17, 18]. E.g. there are several works [19] that show the average delay for some cloud gaming systems (between 135–240 ms. in some cases and between 400–500 ms. in others), but some well-known cloud providers provide quite higher latency values (e.g. measures shown in [17] range between 2.52 and 8.59 s).

Several measurements in IoT systems show that their latency is quite lower. E.g. In [20], all topologies measured have lower average latency than 400 ms, and in [21], authors measured median end-to-end latency between 500–700 ms. So we can consider them as the worst values. Moreover, it is expected that 5G will benefit IoT since its purpose is to provide an average latency of 1 ms.

In order to compare the latency time in MCloud IoT and regular Cloud for IoT, for comparison purposes, we split the latency of the Round Trip Time (RTT) as the Network Delay (ND) plus the Processing Delay (PD). Other delays (OD) like data gathering delay and frame transmission delay are equal in both cases or close to zero. The equation is as follows:

$$RTT = ND + PD + OD \qquad (1)$$

Although the processing delay is lower in the servers provided by a cloud computing service provider than in the mobile devices, we observe in works shown before that the difference of the network latency is quite higher. Taking into account the processing delay at different bandwidths given by Cisco at [22], servers process the information in an order of microseconds, while mobile devices process the data in an order of milliseconds. Figure 5 shows the latency of the compared systems when the packet sizes have 64 Bytes (there are few gathered IoT data per second). We can see that regular Cloud for IoT have higher values than MCloud for IoT. Just the best case of regular cloud for IoT has RTT values in the range of MCloud IoT.

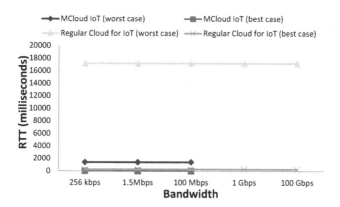

Fig. 5. Latency comparison between MCloud IoT and regular cloud for IoT

5   Conclusion

The integration of IOT and cloud computing is an opening research line and there are some good architectures had been proposed for Cloud IoT. However, they didn't meet the future needs for ubiquitous computing. This paper propose a new MCloud IoT architecture where the things are the mobile devices with ability to build their private cloud taking advantages from their available resources to overcome IOT traditional limitations. Additionally, they are capable to communicate with the neighbor clouds and/or clouds in the internet.

To Implement this architecture a communication protocol will be needed and it is planned as our next step toward illustrate the data flow in the architecture. Combined together with the new architecture the heterogeneity issues will be overcome to enable easy delivering of IOT applications and services in certain scenarios. Moreover, in future work we will add certificate algorithms to secure the system [23, 24].

References

1. Parwekar, P.: From internet of things towards cloud of things. In: 2nd International Conference on Computer and Communication Technology, Allahabad, pp. 329–333 (2011)
2. Botta, A., de Donato, W., Persico, V., Pescapé, A.: Integration of cloud computing and internet of things: a survey. Future Gener. Comput. Syst. **56**, 684–700 (2016)
3. Aazam, M., Khan, I., Alsaffar, A.A., Huh, E.N.: Cloud of things: integrating internet of things and cloud computing and the issues involved. In: 11th International Bhurban Conference on Applied Sciences and Technology (IBCAST), Islamabad, Pakistan, 14th–18th January 2014, pp. 414–419 (2014)
4. Zanoon, N., Rawshdeh, D.: STASR a new task scheduling algorithm for cloud environment. Netw. Protoc. Algorithms **7**(2), 81–95 (2015)
5. Garcia-Pineda, M., Felici-Castell, S., Segura-Garcia, J.: Using factor analysis techniques to find out objective video quality metrics for live video streaming over cloud mobile media services. Netw. Protoc. Algorithms **8**(1), 126–147 (2016)
6. Aazam, M., Huh, E.N.: Fog computing and smart gateway based communication for cloud of things. In: International Conference on Future Internet of Things and Cloud (FiCloud 2014), Barcelona, pp. 464–470 (2014)
7. Abdelwahab, S., Hamdaoui, B., Guizani, M., Znati, T.: Cloud of things for sensing as a service: sensing resource discovery and virtualization. In: IEEE Global Communications Conference (GLOBECOM), San Diego, CA, pp. 1–7 (2015)
8. Bonomi, F., Milito, R., Zhu, J., Addepalli, S.: Fog computing and its role in the internet of things. In: 1st Workshop on Mobile Cloud Computing (MCC 2012), August 2012, pp. 13–16 (2012)
9. Distefano, S., Merlino, G., Puliafito, A.: Towards the cloud of things sensing and actuation as a service, a key enabler for a new cloud paradigm. In: 2013 Eighth International Conference on P2P, Parallel, Grid, Cloud and Internet Computing (3PGCIC), Compiegne, pp. 60–67 (2013)
10. Sotiriadis, S., Bessis, N., Asimakopoulou, E., Mustafee, N.: Towards simulating the internet of things. In: 28th International Conference on Advanced Information Networking and Applications Workshops (WAINA 2014), pp. 444–448 (2014)
11. Liu, Y., Dong, B., Guo, B., et al.: Combination of cloud computing and internet of things (IOT) in medical monitoring systems. Int. J. Hybrid Inf. Technol. **8**, 367–376 (2015)
12. Aazam, M., Hung, P.P., Huh, E.N.: Smart gateway based communication for cloud of things. In: 2014 IEEE Ninth International Conference on Intelligent Sensors, Sensor Networks and Information Processing (ISSNIP), Singapore, pp. 1–6 (2014)
13. Alohali, B., Merabti, M., Kifayat, K.: A secure scheme for a smart house based on Cloud of Things (CoT). In: 6th Computer Science and Electronic Engineering Conference (CEEC 2014), Colchester, Essex, UK, pp. 115–120 (2014)
14. Cai, H., Gu, Y., Vasilakos, A., Xu, B., Zhou, J.: Model-driven development patterns for mobile services in cloud of things. IEEE Trans. Cloud Comput. (In press)
15. Blair, G., Schmidt, D., Taconet, C.: Middleware for Internet distribution in the context of cloud computing and the internet of things. Ann. Telecommun.-annales des télécommunications **71**(3), 87–92 (2016)
16. Macias, E., Suarez, A., Lloret, J.: Mobile sensing systems. Sensors **13**(12), 17292–17321 (2013)
17. Strom, D., van der Zwet, J.F.: Truth and lies about latency in the cloud. Interxion white paper. http://www.interxion.com/globalassets/documents/whitepapers-and-pdfs/cloud/WP_TRUTHANDLIES_en_0715.pdf

18. Arista: Architecting low latency cloud networks, White Paper. https://www.arista.com/assets/data/pdf/CloudNetworkLatency.pdf

19. Chen, K.T., Chang, Y.C., Tseng, P.H., Huang, C.Y., Lei, C.L.: Measuring the latency of cloud gaming systems. In: 19th ACM international conference on Multimedia (MM 2011), Scottsdale, Arizona, USA, 28 November–1 December 2011

20. Kruger, C.P., Hancke, G.P.: Implementing the internet of things vision in industrial wireless sensor networks. In: 12th IEEE International Conference on Industrial Informatics (INDIN 2014), Porto Alegre, Brazil, 27–30 July 2014

21. Shukla, A., Simmhan, Y.: Benchmarking Distributed Stream Processing Platforms for IoT Applications. arXiv:1606.07621v2. 26 July 2016

22. Cisco Systems Inc.: Design best practices for latency optimization, White Paper. https://www.cisco.com/application/pdf/en/us/guest/netsol/ns407/c654/ccmigration_09186a008091d542.pdf

23. Kim, S.: Game based certificate revocation algorithm for internet of things security problems. Ad Hoc Sens. Wirel. Netw. **32**(3–4), 319–336 (2016)

24. Lloret, J., Sendra, S., Jimenez, J.M., Parra, L.: Providing security and fault tolerance in P2P connections between clouds for mHealth services. Peer-to-Peer Netw. Appl. **9**, 5876–5893 (2016). doi:10.1007/s12083-015-0378-3

CoAP-Based Request-Response Interaction Model for the Internet of Things

Fazlullah Khan[1]([✉]), Izaz ur Rahman[1], Mukhtaj Khan[1], Nadeem Iqbal[1],
and Muhammad Alam[2]

[1] Abdul Wali Khan University Mardan, Mardan, Pakistan
{fazlullah,izaz,mukhtaj.khan,nikhan}@awkum.edu.pk
[2] Instituto de Telecomunicações, Aveiro, Portugal
alam@av.it.pt

Abstract. The Internet of Things (IoT) is a broad vision that incorporate real-wold devices from everyday life. These objects coordinate
with each other to share the information gathered from phenomena of
interest. IoT is a broad term and has attain popularity with the integration of Cloud Computing and Big Data. The partnership among these
technologies is revolutionizing the world in which we live and interact
with different devices. On the down side, there are lot of speculations
and forecasts about the scale of IoT products expected to be available
in the market. Most of the products are vendor-specific and as such are
not interoperable. They lack a unified standard and are not compatible
with each other. Another major issue with these products is the lack
of secured features. Albeit, IoT devices are resource-rich, however, they
are not capable to communicate in absence of embedded sensor nodes.
The presence of resource-constrained sensors in the core of each IoT
device make it resource-starving and as such require extremely lightweight but secured algorithms to combat various attacks and malevolent entities from spreading their malicious data. In this paper we aim
to propose an extremely lightweight mutual handshaking algorithm for
authentication. The proposed scheme verifies the identity of each participating device because establishing communication. Our scheme is based
on client-server interaction model using Constrained Application Protocol (CoAP). A 4-byte header, extremely lightweight parsing complexity
and JSON based payload encryption make it a lightweight scheme for
IoT objects. The proposed scheme can be used as an alternative to DTLS
schemes, the one common nowadays for IoT objects.

Keywords: Internet of Things · Constrained application protocol ·
Mutual authentication · Resource-observation

1 Introduction

Technological advances in Micro-Electro-Mechanical- Systems (MEMS) and
wireless communication has formed a solid foundation for sensor-embedded

© ICST Institute for Computer Sciences, Social Informatics and Telecommunications Engineering 2017
J. Ferreira and M. Alam (Eds.): Future 5V 2016, LNICST 185, pp. 146–156, 2017.
DOI: 10.1007/978-3-319-51207-5_14

Internet of Things (IoT) [2]. The basic aim of IoT is to incorporate real-world physical objects by using unique addressing schemes [1]. CISCO has estimated that the number of such objects interconnected with each other and with Internet will surpass 50 billion by 2020[1]. These objects would be enabled to capture, compute and control the events, also known as phenomena of interest, occurring in the real-world [3]. Eventually, this fascinating concept of IoT will lead us to IoE, i.e., Internet of Everything, in which data, systems, objects and interacting processes will be part of it.

Integration of physical objects with the Internet is a challenging task. Security provisioning, an issue faced by the physical devices is of particular concern. This is because each physical object connecting with the Internet has distinguishing features and requirements. Without identity verification, a malevolent entity can easily gain access to a network and perform various malicious and harmful activities. Such malicious activities may include conveying falsified health readings to the doctors residing in distant locations, activation of fake fire alarms in an organization are few of the example in this context. Although, these security threats are highly vulnerable in nature and behavioral, however, very little have been done to secure the inter-connecting physical objects and their end-products. Because of that, the end-products of IoT available in the market are prone to a wide range of security breaches. As a result, Internet of Things (IoT) will eventually leads us to IoV, i.e., Internet of Vulnerabilities.

This paper aims to address the above issues by designing an extremely lightweight but highly secured and robust authentication scheme. The goal of our proposed scheme is to verify the identities of communicating objects in the IoT paradigm. Our proposed scheme works on the application layer of any physical object and uses a well-known Internet of Things (IoT) protocol, known as Constrained Application Protocol (CoAP) [7] for the network operation. This paper has two major objectives. It first authenticates the identities of the physical objects inter-connecting with each other. Each physical object, in a role of a client, communicate with a given server for authentication. Authentication is mutual because both the client and server mutually authenticate the identities of each other. Unless both entities are authenticated, a connection, i.e., a session will not establish between them. Once authentication is successful, the clients are eligible to observe the resources at a given server. Each server resides a set of resources which can only be observe by a legitimate client, i.e., the one which has been authenticated successfully. Each client has the ability to specify certain conditions to the server for resource observation. Such conditional specification not only enable a client but also the server to conserve their limited resources. Resources are only observe once the condition for observing a resource are fulfilled at the server end. These two objectives are vital for any robust and secured communication system. Conditional resource observation is highly essential in these networks because each object has its own requirement for data observation, data rate, memory availability and sleeping schedule. The latter attribute is because of the embedded sensor node at the core of each physical object. In

---

[1] http://www.cisco.com/web/solutions/trends/iot/indepth.html.

our proposed scheme, data flow between a client and a server only commences once a successful session is authorized.

The rest of this paper is organized as follows. In Sect. 2, related work is presented followed by the proposed scheme in Sect. 3. Experimental work and analysis are provided in Sect. 4. The paper in concluded and future research directions and gaps are discussed in Sect. 5.

2   Related Work

In this section, we present related work on mutual authentication and resource observation in the Internet of Things (IoT) paradigm. Today, the Internet of present is mainly based on REpresentational State Transfer (REST) architecture. The said architecture uses HTTP protocol [4] for its operation. HTTP, on the other hand, is a resource-consuming protocol which require ample amount of storage and computational resources. The real-world physical objects in an IoT paradigm are highly resource-starving due to the underlying embedded sensor nodes and as such lack the support for HTTP protocol. For the provisioning of RESTful services in any resource-constrained network, Internet Engineering Task Force (IETF) has come with an extremely lightweight protocol, known as Constrained Application Protocol (CoAP). This protocol is a lightweight version of HTTP, however, it is not an alternative of the latter. Our previous work on secure communication and architecture for wireless sensor networks can be studied in [5, 6].

CoAP was designed in view of limited resources of the objects. This protocol allows the exchange of messages between resource-starving physical objects over resource-limited communication networks [7]. In the communication context, resource-starving objects are miniature devices which lack the support for processing speed, power, storage, available bandwidth and data rate. Such devices are often built using an 8-bit or 16-bit micro-controllers. In some case, the micro-controllers have an upper bound of 32-bit. Unlike conventional networks, the resource-limited networks lack the support for a fully functional TCP/IP stack. IPv6 over Low-power Wireless Personal Area Network (6LoWPAN) is a well-known example of such networks. Instead of TCP, CoAP uses UDP at transport layer for flow control and session initiation and work alike HTTP to match requests with corresponding responses. Similar to web, IP addresses and port numbers are used to locate a resource residing on a given serve. Various RESTful URIs are used to provide access to the resources. Methods such as GET, POST, DELETE and PUT are used in similar fashion to HTTP. CoAP is not a replacement of HTTP protocol, however, it uses a small subset of commands and context of HTTP to optimize for Machine-to-Machine (M2M) exchanges. CoAP can be considered as a method for accessing and invoking various RESTful services exposed by physical objects, also known as Things, over a physical network. CoAP supports four different types of messages and their specification are defined in the CoAP-draft [7].

1. **CON**: CON represents a confirmable message which requires a valid response. The said response can either be positive or a negative acknowledgement. If in case, an acknowledgment is not received by the sender, the request is re-transmitted until all such attempts for transmission are exhausted. The re-transmissions attempts increase in a non-linear, exponential fashion.
2. **NON**: NON represents a non-confirmable message and is used for unreliable transmission such as a request for sensor readings which are observed periodically. In such transmission, if one reading value is missed, there is little impact on the overall reading. NON messages are not acknowledged and the response is mostly NON as well.
3. **ACK**: ACK represents a valid acknowledgement and is either piggybacked in the response or send as a separate message. ACK is sent in response to a CON message and contains information about an observed data. If ACK is lost, the response need to be send again with the same ACK by the server.
4. **RST**: RST generally represents a negative acknowledgement and is used when the server wakes up from sleep mode and lose the context of the previous state.

In the Internet of Things paradigm, the resources residing on a given server need to be observed in a secured manner. A wide range of security challenges faced by IP-enabled real-world physical objects are highlighted in [8]. In view of these challenges, extremely lightweight, robust and secured protocols need to be designed to meet the requirement of resource-starving sensor-embedded objects communicating over resource-constrained networks. Despite the presence of sensor nodes at the core of each object, the wide-range of security protocols available in literature for Wireless Sensor Networks (WSNs) are not applicable to these objects [9]. This is due to the fact that sensor-embedded physical objects have their own unique attributes and characteristics and as such does not suit the available protocols for WSNs. Any designed protocol for IoT need to be lightweight as well in view of the underlying resource-constrained sensor nodes in each object.

An RSA-based encryption algorithm for the IoT objects was proposed in [10]. The proposed scheme used a pair-wise key, i.e., a public key and a private key. The proposed scheme is, however, highly resource-consuming and require heavyweight and resource-intensive cryptographic suites. As a result, it does not meet the demands of resource-starving objects. A server-based certificate validation protocol was proposed in [11]. The said protocol enables one or more clients to delegate certificate validation to an entrusted server. However, the proposed protocol increases communication overhead and as such does not fit to the requirement of resource-constrained objects of an IoT. Certificate validation and PKI are well-known cryptographic and authentication schemes in the Internet. However, for the IoT, these schemes are highly complex in terms of computation, storage and as such require proper configuration to suit the objects interacting in an IoT requirement. Implementation of key-pair approaches restricts miniature sensor-embedded objects from utilizing these schemes. Data-gram Transport Layer Security protocol (DTLS) is an obvious choice for IoT objects because CoAP protocol uses UDP as a default and resource-saving scheme [12].

However, DTLS with full PKI is not an optimal choice for IoT objects. A symmetric key encryption scheme was proposed by [13] for authentication. The proposed scheme uses a single pre-shared secret for establishing a communication session. Although, the proposed scheme reduces energy consumption and computational resources, however, it has not been validated via experimental results.

In light of the aforementioned discussion, we propose an extremely lightweight authentication approach which uses a single key for authentication. Our scheme incurs very small overhead and is sufficiently simple in terms of computation and resource utilization. A 4-way handshaking approach is adopted to authenticate the interested clients and servers. Upon successful authentication, each client registers itself with the server to observe a resource, temperature readings in our case. Malicious clients are prevented to observer resources and from establishing connections with a server. Each client is restricted only to a single connection for fair utilization of resources.

3   CoAP-Based Request-Response Interaction Algorithm

In this section, a brief overview of our CoAP-based 4-way handshake mechanism for authentication is presented. It is important to note that our scheme and CoAP are not two separate protocols. Instead, we use our own security patch embedded in CoAP for authentication purpose to tackle various attacks.

Similar to any other communication network, resource preservation is a challenging task and is of utmost importance. Data fabrication by malicious entities and its spread over a network will jeopardize the traffic flow of the whole network. As a result of data fabrication, each object in the network will end up with large number of copies of the malicious data. Not only the client, but the server is also vulnerable to be compromised by a malevolent entity. Therefore, it is mandatory to authenticate the integrity and identity of both the parties, i.e., the client and the server.

In light of the above discussion, we have proposed a lightweight algorithm which uses the underlying operational model of CoAP. The proposed scheme can be use as a lightweight alternative to DTLS because its simple to implement, flexible in terms of complexity and infrastructure. Each client and a server challenge for mutual authentication using the exchange of four simple handshake messages. Each message comprise of 256 bits. The only exception is the first message, i.e., initial session initiation request. The small size of messages incur small overhead during the authentication procedure and causes less burden to the IoT objects. We have used Advanced Encryption Standard 128 bits, i.e., AES-128 for authentication and encryption. The four phases of our authentication schemes are *Session Negotiation, Server Challenge, Client Challenge and Response*, and *Server Response*.

Before session negotiation phase, each client shares with a server a 128 bit preshared secret Y_i. This is a pre-requisite phase which takes place well before any authentication commences. Each object has a unique identity (ID) associated with it which enables a server to look-up for that ID in its table. Y_i is known only to the client and the server as it is pre-distributed before any authentication commences. The first phase, i.e., the session initiation or negotiation, is validated once a match is found by the server in the table. If the ID of a client is not present, the server will not proceed to session initiation. Figure 1 shows the Key-ID pairs in the table maintained by the server.

Device ID(i)	1	2	3	4	5	6	......	n
Pre-shared Key Y_i	Y_1	Y_2	Y_3	Y_4	Y_5	Y_6	......	Y_n

Fig. 1. Pre-shared secrets and IDs in server table

ID matching with the table only allows the server and a client to communicate with each other to exchange a session key. The actual authentication is completed only using the four handshake messages as shown in Fig. 2. The handshake procedure is explained in details in the following four phases.

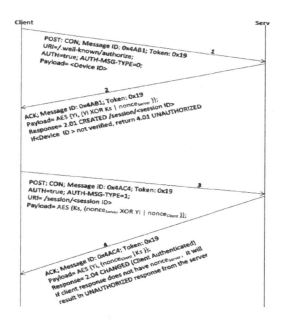

Fig. 2. Four-way authentication handshake

Session Negotiation. After a successful match of pre-shared secret, the actual authentication starts using the four-way handshake mechanism, i.e., four phases. In the first phase, a session is negotiated between the client and a server. Each client sends a request message to the server. This message is CON and the method is POST. The purpose of this message is to create a resource at the server. Each message contains a token which is used to correlate the CON request with a matching response (ACK). Also, each message has its own ID, to uniquely identify it. Each client has the ability to maintain and monitor a buffer, which contains all transmitted request, i.e., CON messages to the server. If an ACK is not received within the specified duration, the CON message is re-transmitted. A CON message is also re-transmitted if the message timeouts. As shown in the Fig. 2, the session negotiation message also carries two options, i.e., Auth and the Auth-Msg-Type. URI is also present in the message which directs the client request towards a given resource. Here, in Fig. 2, /authorize is a resource residing at the server. In our scheme, the resource is temperature readings captured by a given server. The value of Auth = true, Auth-Msg-Type = 0 and /authorize is an indication to a server that the request is for session negotiation.

Server Challenge. Upon reception of session negotiation request at the server end, Object ID, is retrieved from the message payload. It enables a server to find a matching Y_i which is associated with a given client. If a match is found, then the server responds back with a payload which is encrypted using Advanced Encryption Algorithm (AES-128 bit). A pseudo-random number, a nonce (nonce$_{Server}$), and potential session key K_s are generated. All these parameters are of 128 bits. The nonce, on the other hand, is used only once by the server in the entire authentication mechanism. Using these parameters, the server generates an encrypted payload. An XOR operation is performed on K_s and Y_i. Then, the resultant is appended with nonce$_{Server}$ and is encrypted with Y_i. All these steps formulate Eq. 1.

$$E_{payload} = AES\{Y_i, (Y_i \ XOR \ K_s | nonce_{Server})\} \qquad (1)$$

In this equation, $E_{payload}$ is the resultant encrypted payload generated by the server as a challenge which need to be decrypted by a given client. Only a legitimate client can decrypt the payload by using the appropriate pre-shared secret.

Client Response and Challenge. When the client receive the encrypted payload, i.e., the result of server challenge, it needs to decrypt the said payload for the retrieval of K_s. If successful, the client will have the original K_s and nonce$_{Server}$. To decrypt the payload of server challenge, each client uses its unique Y_i, which is known only to a given client. Successful decryption of server challenge means that the given client has been able to authenticate itself. As our proposed scheme is based on mutual authentication, hence, the server also need to be authenticated. To do so, each client generates its own challenge, an encrypted payload, similar to the server challenge. For this, each client generates an encrypted payload of its own by using XOR operation as before. Nonce$_{Server}$ and Y_i are used as the two parameters for an XOR operation. The resultant of this operation is then appended to nonce$_{client}$ and encrypted with K_s. The detailed operation is shown in Eq. 2.

$$E_{payload} = AES\{K_s, (nonce_{Server} \ XOR \ Y_i | nonce_{Client})\} \qquad (2)$$

Here, $E_{paylaod}$ is an encrypted payload generated by the client and nonce$_{Client}$ is a pseudo-random number, similar to nonce$_{Server}$, however, it is generated by the client. Similar to nonce$_{Server}$, nonce$_{Client}$ is generated and used only once in an authentication process. During this phase, Auth = *true* and Auth-Msg-Type = 1 literally means that the server should realize that this request is different than session negotiation and it means that the encrypted payload in a server challenge was successfully deciphered by the client. At this point, the potential session key, K$_s$ has been securely transmitted to the given client.

Server Response. In this final phase, the server retrieves the encrypted payload from the client challenge. Upon observing nonce$_{Server}$ in a client response, the server knows that the given client has been successful to authenticate itself. Now, the server also needs to decrypt the payload by retrieving nonce$_{Client}$ from it. Upon successful decryption, the server creates a payload and embed the nonce$_{Client}$ in it and appends K$_s$ with it. This encrypted payload is encrypted with Y$_i$ as shown in Eq. 3.

$$E_{SP} = AES\{Y_i, (nonce_{Client}|K_s)\} \tag{3}$$

The client has already authenticated itself. So, the server changes the status of the temperature resource to *Authenticated*. Now, the encrypted payload is being transmitted to the given client. When the client receives it, it decryptes it and observe nonce$_{Client}$ in it. By observing nonce$_{Client}$ in the encrypted payload, the client realizes that the server has also authenticated itself. With this phase, both parties are mutually authenticated, are they are now ready to exchange data between themselves.

In our proposed scheme, we have created our own Options similar to the one CoAP uses. The formats of these two options, Auth and Auth-Msg-Type, are shown below in Fig. 3.

No.	C	U	N	Name	Format	Length	Default
TBD	X	X	-	Auth	empty	0	(none)
TBD	X	X	-	Auth-Msg-Type	uint	1	(none)

Fig. 3. Option formats

At this point of time, our proposed scheme only emphasis on authentication. We are still working on the actual exchange of data. For this we will specify various conditions for the exchange of data. Without successful authentication, any exchange of data is meaningless.

4   Experimental Evaluation

In this section, we have discussed the initial evaluation of our scheme. Before the initiation of communication, client-server authenticates each other by validating their IDs. For authentication and conditional resource observation, we have applied an open source library, i.e. CoAPSharp. This library consist of basic CoAP protocol and offers normal resource observation. Therefore, we have modified the existing protocol with our authentication scheme and application specific conditional options.

In the implementation phase, we did our evaluations on the emulators first, and then confirmed and implemented on the NetDuino Plus 2 boards. A temperature sensor, Dellas DS1820 was embedded on the NetDuino Plus 2 Board. The NetDuino board in the role of a server provides conditional specific resources to four different clients in our proposed scheme. Each NetDuino Plus 2 board control an application, as discussed in the previous section. Hence, our test-bed is made-up of a total of five boards, a server and four client NetDuino boards.

Prior to setup a conditional resource observation relationship the client-server must authenticate each other. Figure 4 shows a successful authentication of this communication. In this figure, the server key is the potential session key which needs to be securely and successfully transmitted to each client. Upon successful decryption, the authentication process is completed. Here, the client key is the pre-shared secret key associated with each client.

```
The thread '<No Name>' (0x2) has exited with code 0 (0x0).
[SERVER] Started.
[SERVER] Key: 4F9DB1949D924031-8C77BE06276ECB25 Nonce1: 2F2515012EDB8CE0-5A02705303A1544C
Nonce2:
[CLIENT] Key: 16BBE8D16B4C00F8-3143F1D60DA5E97D Nonce1: 2F2515012EDB8CE0-5A02705303A1544C
Nonce2: 5E10A9012748BDDA-3FFDCFF6128F4056
[CLIENT] Replying to server challenge...
[SERVER] Access granted to client 16BBE8D16B4C00F8-3143F1D60DA5E97D
[SERVER] Key: 4F9DB1949D924031-8C77BE06276ECB25 Nonce1: 2F2515012EDB8CE0-5A02705303A1544C
Nonce2: 5E10A9012748BDDA-3FFDCFF6128F4056
```

Fig. 4. Successful authentication response

Figure 5 shows an unsuccessful authentication response. Here, the client is unable to decrypt the session key. Therefore, the client is banned from registering with the server for the resource observation. Failure to decrypt the session key eliminates various types of attacks in an IoT environment.

```
'Microsoft.SPOT.Emulator.Sample.SampleEmulator.exe' (Managed): Loaded
'C:\Users\mian\Desktop\sources-20140528\sources\CoAPTest-Server\bin\Debug\le\CoAPTest-
Server.exe', Symbols loaded.
The thread '<No Name>' (0x2) has exited with code 0 (0x0).
[CLIENT] Started.
The thread '<No Name>' (0x2) has exited with code 0 (0x0).
[SERVER] Started.
[SERVER] Key: 4F9DB1949D924031-8C77BE06276ECB25 Nonce1: 4BAFCDE9430E5773-24E5095A66614BF17
Nonce2:
[CLIENT] Key: 6619DB083FA7049A-70F225566ED3847A Nonce1: 4BAFCDE9430E5773-24E5095A66614BF17
Nonce2: 6DFB243F1F64BE50-142C6CFE3382DAAF
[CLIENT] Replying to server challenge...
[CLIENT] Resource access denied.
```

Fig. 5. Unsuccessful authentication response

In Fig. 6, the status of various physical devices registered with the server for conditional resource observation is depicted.

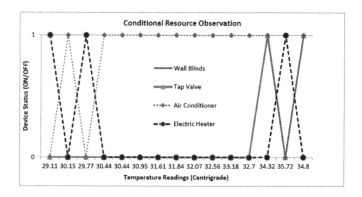

Fig. 6. Conditional resource observation

Here, each device relies on the temperature readings of the server. We have different and specific condition for the announcement of various messages to the server. Each device remains in a particular state (ON/OFF) and switches its state once a particular condition is fulfilled. Different conditions specified for our experimental results are already explained in the previous section. Here, 0 represents OFF and 1 represents ON state.

In the above figure, we have provided the preliminary results. Currently, we are conducting extensive mathematical and experimental evaluation of our proposed scheme against the DTLS and PKI in terms of various performance metrics like the latency, packet loss, throughput, data rate, and average battery power consumption.

5   Conclusion

In an Internet of Things (IoT), not much efforts have been made for securing the IoT products available in the market. A large number of such products are reaching to the market, however, most of these products lack security features. Because, each object of an IoT has its own peculiar characters and has different attributes, the existing secured solutions available for the Internet are not feasible for apply to them. The presence of embedded sensors in each object does not mean that secured solution for WSNs are applicable to these networks because, of their own unique and distinguishing underlying hardware and software prototypes.

Our algorithm is highly efficient against key fabrication, resource exhaustion, eavesdropping and DoS attacks. However, it may not be efficient against Sybil attack [14]. But again, no secured solution in research can tackle all type of attacks. Despite the buzz and hype surrounding around Internet of Things, secured features will always remain a major concern for their products and objects due to their unique features and foremost we do not know what a real-world object will behave when it is connected with Internet. Such challenges encourage academia and industry to explore in-depth and come up with various innovative solutions to tackle security loophole and vulnerabilities faced by these objects.

References

1. Atzori, L., Iera, A., Morabito, G.: The internet of things: a survey. Comput. Netw. **54**(15), 2787–2805 (2010)
2. Alam, M., Ferreira, J., Fonseca, J. (eds.): Intelligent Transportation Systems: Dependable Vehicular Communications for improved road safety, vol. 52. Springer, Heidelberg (2016). ISSN 2198-4128
3. Bormann, C., Castellani, A., Shelby, Z.: CoAP: an application protocol for billions of tiny internet nodes. IEEE Internet Comput. **16**(2), 62–67 (2012)
4. Fielding, R.T., Taylor, R.N.: Principled design of the modern web architecture. ACM Trans. Internet Technol. (TOIT) **2**(2), 115–150 (2002)
5. Khan, F.: Secure communication and routing architecture in wireless sensor networks. In: IEEE 3rd Global Conference on Consumer Electronics (GCCE), pp. 647–650. IEEE (2014)
6. Khan, F., Bashir, F., Nakagawa, K.: Dual head clustering scheme in wireless sensor networks. In: International Conference on Emerging Technologies (ICET), pp. 1–5. IEEE (2012)
7. Shelby, Z., Hartke, K., Bormann, C., Frank, B.: Constrained Application Protocol (CoAP), draft-ietf-core-coap-13. The Internet Engineering Task Force-IETF, Orlando (2012)
8. Heer, T., Garcia-Morchon, O., Hummen, R., Keoh, S.L., Kumar, S.S., Wehrle, K.: Security challenges in the IP-based internet of things. Wirel. Pers. Commun. **61**(3), 527–542 (2011)
9. Jan, M.A., Nanda, P., He, X., Liu, R.P.: PASCCC: priority-based application-specific congestion control clustering protocol. Comput. Netw. **74**, 92–102 (2014)
10. Milanov, E.: The RSA algorithm, June 2009
11. Freeman, T., Malpani, A., Cooper, D., Housley, R.: Server-based certificate validation protocol (SCVP) (2007)
12. Hartke, K., Bergmann, O.: Datagram transport layer security in constrained environments (2012)
13. Jan, M.A., Nanda, P., He, X., Tan, Z., Liu, R.P.: A robust authentication scheme for observing resources in the Internet of Things environment. In: IEEE 13th International Conference on Trust, Security and Privacy in Computing and Communications, pp. 205–211. IEEE (2014)
14. Jan, M.A., Nanda, P., He, X., Liu, R.P.: A sybil attack detection scheme for a forest wildfire monitoring application. Future Gener. Comput. Syst. (2016)

Performance Analysis of Vehicular Adhoc Network Using Different Highway Traffic Scenarios in Cloud Computing

Nigar Fida[1], Fazlullah Khan[1(✉)], Mian Ahmad Jan[1], and Zahid Khan[2]

[1] Department of Computer Science, Abdul Wali Khan University Mardan,
Mardan, Pakistan
nigar.fida@yahoo.com, {fazlullah,mianjan}@awkum.edu.pk
[2] Institute of Mobile Communication, Southwest Jaiotong University, Chengdu,
People's Republic of China
zahid@my.swjtu.edu.cn

Abstract. Vehicular Ad-hoc Networks (VANETs) combine intelligent vehicles on highways aim to solve many transportation problems. The performance of VANETs is affected by many parameters due to highly dynamic structure. We assessed the performance of VANETs over different highway's scenarios and investigated that under which circumstances the performance will be better and vice versa. We adopted our experiments in infrastructure environment, where the road side units (RSUs) are connected with cloud server. The RSU periodically gathers spatial-temporal information and upload it to cloud, which could help the drivers to predict the status of road before journey. The experiments carried on two types of highway's scenarios: varying vehicles densities and simulation time. The simulation result shows that selected performance metrics (throughput, E2E delay and packet loss) greatly affect in both scenarios. The simulation time within the interval 200 to 500 is an optimal choice during simulation experiments. The throughput and packet loss increases with increase in vehicle density. The end-to-end delay has an inverse relation with vehicle density. The highway scenarios are generated by SUMO and the actual simulation is done by NS2.

Keywords: Cloud computing · Network simulator · Vehicle density · VANET · Performance analysis · Throughput · Packet loss · End-to-end delay

1 Introduction

The adoption of Vehicular Ad-hoc Networks (VANETs) across industry has increased due to advancements in ad-hoc wireless technology. Vehicular ad hoc networks (VANETs) are classified as an application of mobile ad hoc network (MANET) that has the potential to solve many Intelligent Transportation System (ITS) problems. Recently VANETs have emerged to turn the attention of researchers in the field of wireless and mobile communications. VANET differs from MANET by architecture, challenges, characteristics and applications. A VANET has some particular features despite being a special case of MANET and presenting some similar characteristics as well [1].

© ICST Institute for Computer Sciences, Social Informatics and Telecommunications Engineering 2017
J. Ferreira and M. Alam (Eds.): Future 5V 2016, LNICST 185, pp. 157–166, 2017.
DOI: 10.1007/978-3-319-51207-5_15

VANETs can be widely used in safety, traffic information, and other commercial applications such as a road side restaurant advertisement, digital entertainment, etc. [2, 3]. Based on VANETS, Intelligent Transportation System (ITS) [4] is very efficient to improve safety, and reduce transportation times and fuel consumption. In literature, VANETs have two operational modes such as V2I (vehicle to infrastructure) and V2V (vehicle to vehicle). In V2I mode, the vehicles communicate through a central base station called Road Side Unite (RSU) which collects data periodically and send it to the cloud server for traffic management, road safety and decision making. While in V2V, vehicles are independently communicating with each other. Figure 1 shows the architecture of V2V and V2I.

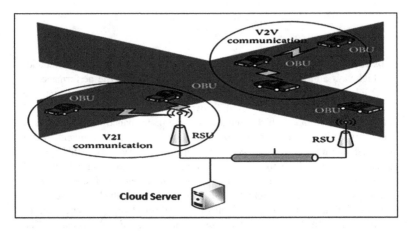

Fig. 1. V2V and V2I operation structure

The high mobile structure of VANET is a challenging task. The existing mobility models are not too dynamic to maintain the highway topology. To get control over the randomize topology structure of VANETs, Vehicle should be enough intelligent to maintain its status. For a consistent and efficient VANET topology, we need two things, a good mobility model that keep track of each vehicle at each moment and second a strong routing algorithm. Routing is the selection of optimal routes for forwarding packets [5]. In this work, we considered a highway of 1 km long for vehicles having RSUs which are further connected with cloud server. The highway mobility model is designed by a prominent simulation framework called Sumo (Simulation of urban mobility) [6]. The highway mobility model is converted to NS2 executable form by MOVE (Mobility Model Generator for Vehicular Networks) [7].

The routing algorithm used in highway simulation is Ad-hoc on Demand Distance Vector (AODV). AODV is a reactive protocol, which discovers routes on demand. AODV can be used in many types' scenarios such as unicast, multicast and broadcast [8]. The different types of control messages used in AODV are Route Request (RREQ), Route Reply (RREP) and Route Error (RERR) [9]. Ahmad *et al.* [9] assessed the performance of OLSR protocol in MANET using different scenarios and performance parameters. We took their work as a motivation and adopted the same scenario cases and performance parameters in VANET Cloud computing. We assessed the performance of AODV protocol in VANET by varying highway topologies and using the same parameters as that of Ahmad *et al.* [9].

We measured throughput, packet loss and End-to-End delay under different highway scenarios and concluded that the given parameters significantly effect by increasing vehicles density. The simulation time after 500 s doesn't significantly affect the performance of highway's vehicles. The statistics of all performance parameters are uploaded to the cloud for efficient traffic planning.

The rest of the paper is organized as follows: Sect. 2 describes the existing work, Sect. 3 having simulation results and discussion, and finally in Sect. 4, we concluded our work.

2 Literature Overview

Das et al. in [10] evaluated the performance comparison of various adhoc routing protocols i.e. LARI, AODV and DSR in terms of Packet Delivery Ratio for VANETs. The performance of these protocols is studied with varying node speeds and traffic density. As a result, the LARI protocol outperforms than AODV and DSR when the network is sparsely populated. The successful delivery of the message was nearly 99.52% in LARI protocol.

The authors in [11] evaluated the performance of reacting protocols namely, AODV and DSR on highly dense and mobile network based on different parameters i.e. Throughput, End-to-end delay and Packet Delivery Ratio. They compared the performance of their selected protocols with newly developed protocol DYMO.

Their simulation shown, that the overall throughput of DYMO is much better than the other two protocols and also end-to-end delay was lowest, when compared with the other (AODV, DSR). However packet delivery ratio of AODV was better than DYMO and DSR. Perdana *et al.* in [12] evaluated the performance of PUMA routing protocol using Manhattan Mobility Model with effect of Nakagami fading distribution in VANET. The performance was analyzed by varying the traffic parameters in the Manhattan mobility model from low to high traffic condition. At the end, they concluded that Nakagami fading and Manhattan mobility model affected the Quality of Service (QoS) in VANET. In [13] the authors analyzed the performance of AODV and GPSR routing protocols in VANET in different scenarios under different traffic conditions with respect to Packet Delivery Ratio (PDR) and average End-to-End Delay (E2ED). Their simulation results showed that AODV performs better with respect to PDR and GPSR outperforms AODV with respect to E2ED. Also, the performance of both the routing protocols depends on scenario and traffic's type.

Shaheen *et al.* [14] examined two routing protocols (AODV and DSR) over two different scenarios (dense and sparse). Both protocols are measured using different performance metrics such as Packet Delivery ratio (PDR), Throughput, End-to-End Delay. According to their simulation results, AODV outperformed than DSR in case of dense scenario, while the DSR protocol is better than AODV, when the network scenario was sparsely populated. Ahmad *et al.* [9] studies the most prominent and widely used protocol i.e. Optimized Link State Routing (OLSR) for MANETS. Their paper presents the performance evaluation of OLSR protocol for TCP and UDP traffic patterns by varying parameters like node density, node speed and pause time. The performance of OLSR has been assessed in terms of performance metrics, such as packet loss, end-to-end delay and throughput under different network scenarios. The results prove that, TCP performs considerably well in terms of throughput, end to end delay and packet loss in different node density and mobility scenarios, while UDP is better in case of pause time.

In this paper, we considered one kilometer long highway and assessed the performance of ad-hoc on demand distance vector (AODV) using the same parameters used in [10]. Our simulation is adopted in real life highway scenario, where the performance is analyzed under different simulation time and vehicle's density. Also the vehicular cloud computing technology is embedded in experiments to minimize traffic congestion, accidents, travel time and environmental pollution. Further our previous work on performance improvement is available [15–19].

3   Simulation Results and Discussions

This is section having information about the simulation environment, tools and results along with discussion. In short, the simulation is carried out on a long highway of length 1 km having vehicles vary from 10 to 100. The ad-hoc on demand distance vector (AODV) works on the top of each vehicle to trace packet transmission. The vehicles will be equipped with communication, computing and sensing devices, and the universal networks will make the internet available during travelling. Thus, the driving experience will be more enjoyable, comfortable, safe and environmental friendly. The performance parameter for assessment VANETs robustness under varies vehicle densities and simulation times are throughput, packet loss, and end-to-end delay.

3.1   Simulation Environment

The high dynamic nature of VANET faces many challenges in real adaptation. The instance change in topological structure of VANET is a big challenge for all existing mobility model [16]. We used three different tools for our experiments. The combo of Sumo and Move are used for designing an operational highway of ongoing vehicles. On the other hand, NS2 is used for actual performance assessment under different scenarios. All the sensed data of highway's vehicles are collected through RSUs, which are further uploaded to the cloud for driver decision making. In this paper we have used Network Simulator version 2 (NS-2) is one of the prominent simulators used to simulate VANETs routing protocols in different scenarios. Manhattan Grid model is adopted in simulation for the movement of nodes [17]. This mobility model is used by NS2.35 to simulate realistic vehicle movement. The simulation parameters for highway traffic design are given in Table 1. We have used Simulation of Urban Mobility (SUMO) [6] is an open source, highly portable, microscopic road traffic simulation package designed to handle large road networks and the Mobility Model Generator for

Table 1. Highway scenario parameters

Parameter	Value
Simulator	Sumo 0.12.3, Move v2.9
Mobility model	Manhattan grid
Simulation time	200, 400, 600, 800, 1000
Highway length	1 km
Number of junctions	3
Number of nodes	10–100
Vehicle's min speed	100 km/h
Vehicle's max speed	120 km/h
Number of lanes	2
Traffic lights	3

Vehicular Networks (MOVE) was recently introduced to make ease of Sumo usage [7]. It is a simple parser for the SUMO and enhances SUMO's complex configuration with a nice and efficient GUI. The simulation parameters are given in Table 2.

Table 2. Simulation conditions

Simulation conditions	Value
Traffic type	UDP
Number of nodes	100
Routing protocol	AODV
Mac protocol	IEEE 802.11p
Packet size	1000

3.1.1 Network Simulator

Network Simulator (NS-2) is one of the prominent simulators used to simulate VANETs routing protocols in different scenarios. Manhattan Grid model is adopted in simulation for the movement of nodes [17]. This mobility model is used to simulate realistic vehicle movement. The simulation parameters are given in Table 2.

3.2 VANETs Performance Analysis on Highway's Vehicles

In this section, we evaluated the performance VANETs over highway with respect to different vehicle density and Simulation time. The performance is assessed by three performance parameters such as throughput, delay, and packet loss. The simulation environment is considered according to Tables 1 and 2.

3.2.1 Simulation Time Impact on Highway's Vehicles

In order to assess the simulation time impact on highway's vehicles, we kept the speed and vehicle density constant to 100 km/h and 50 Veh/km vehicles respectively.

3.2.1.1 Throughput vs Simulation Time

The line graph in Fig. 2 shows that throughput increase with increase in simulation time. Hence we concluded that for an optimal simulation experiments, we have to keep the simulation time in the interval 500 to 1000 s.

A small simulation time causes unrealistic performance, because all the nodes are not participating in short simulation run. Figure 2 also depicts that after a specific simulation interval, the throughput remains constant as shown in simulation time 800 and 1000 s.

3.2.1.2 End-to-End Delay vs Simulation Time

The end-to-end delay in a short simulation run seems quite long duration as shown in Fig. 2. In short, we can say that, delay and simulation time are indirectly proportion to each other. The end to end delay remains constant after specific simulation run time as shown in Fig. 2.

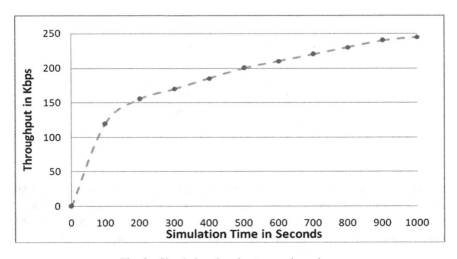

Fig. 2. Simulation time impact on throughput

3.2.1.3 Packet Loss vs Simulation Time

The packet loss has directly proportional relation with simulation time as shown in Fig. 3. But after getting a specific simulation time run, we will experience constant packet loss rate. From Figs. 1, 2, 3, and 4, we concluded that we should keep a moderate simulation time for experiments. Small simulation duration causes to un-realistic results, while too long just waste of time.

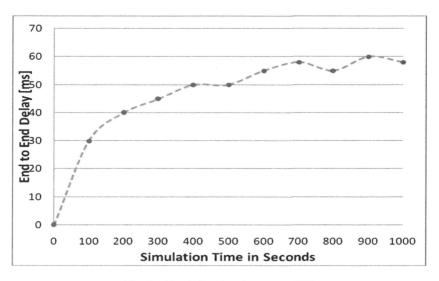

Fig. 3. Simulation time impact on E2E

3.2.2   Vehicle Density Impact on Highway's Vehicle

Vehicle density refers to number of nodes passing through per unit highway's length. In this section, we analyzed the vehicle density impact on highway's vehicles by using throughput, delay and packet loss.

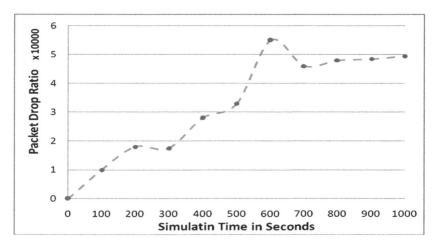

Fig. 4. Simulation time impact on packet loss

3.2.2.1 Throughput vs Vehicle's Density

The average throughput increases with vehicle density as shown in Fig. 5. We kept 10% of the vehicles as traffic agents on highway. Hence with increase in vehicle's density the traffic agents also increases and over all throughput increases.

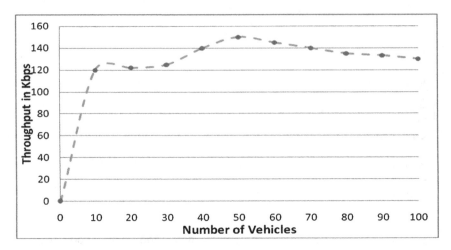

Fig. 5. Vehicle density impact on throughput

3.2.2.2 End-to-End Delay vs Vehicle's Density

The relationship between delay and density is not very clear as depicts in Fig. 6. It has some exponential up and down, which shows that position of communicated nodes also effect the performance.

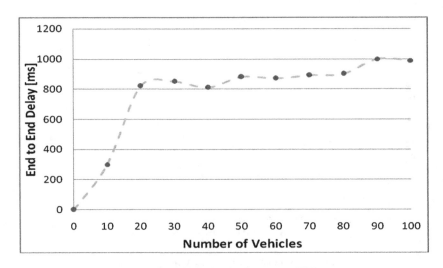

Fig. 6. Vehicle density impact on E2E

3.2.2.3 Packet Loss vs Vehicle's Density

The increase of vehicle's density on highway degrades the performance of packet delivery as shown in Fig. 7. The high packet drop in Fig. 7 is a notification of traffic jam.

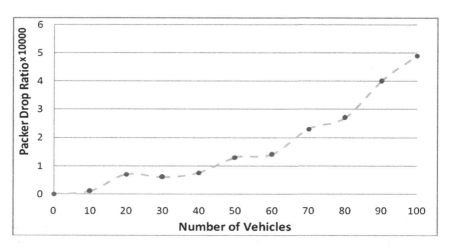

Fig. 7. Vehicle density impact on packet loss

4   Conclusion

In this paper, we analyzed the performance of VANET using different highway's traffic scenarios in cloud computing. From simulation results, we concluded that, Simulation time has significant affect on VANETs performance. A moderate simulation time will be an optimal choice for good results. In a short time interval, all the nodes will not participate in the simulation, which will affect the results. Similarly keeping to much long time has no impact, because after a specific time, the values of all performance parameters become constant. The throughput and packet loss are directly proportional to vehicle density. The throughput and packet loss in a congested condition are significantly more than normal traffic. The relation between End-to-End delay and vehicle density is ambiguous as it has some variability which affects the performance. Our design framework will provide an intermediate platform in the form of cloud server, which could provide the ongoing status of highway.

References

1. Cunha, F., Villas, L., Boukerche, A., Maia, G., Viana, A., Mini, R.A., Loureiro, A.A.: Data communication in VANETs: protocols, applications and challenges. Ad Hoc Netw. **44**, 90–103 (2016)
2. Alam, M., Ferreira, J., Fonseca, J.: Introduction to intelligent transportation system. In: Alam, M., Ferreira, J., Fonseca, J. (eds.) Intelligent Transportation Systems: Dependable Vehicular Communications for Improved Road Safety, vol. 52, pp. 1–17. Springer, New York (2016)
3. Araghi, T.K., Zamani, M., Mnaf, A.B.T.: Performance analysis in reactive routing protocols in wireless mobile ad hoc networks using DSR, AODV and AOMDV. In: 2013 International Conference on Informatics and Creative Multimedia (ICICM). IEEE (2013)

4. Alam, M., Fernandes, B., Silva, L., Khan, A., Ferreira, J.: Implementation and analysis of traffic safety protocols based on ETSI standard. In: IEEE Vehicular Networking Conference (VNC) Kyoto, Japan (2015)
5. Khan, Z., Tayeba, H.F., Awan, I.I., Nawaz, A.: Impact of mobility models over multipath routing protocols. Int. J. Inf. Technol. Comput. Sci. (IJITCS) **19**(1) (2014). http://www.ijitcs. com
6. SUMO - Simulation of Urban MObility. http://sumo.sourceforge.net
7. Bilandi, N., Verma, H.K.: Comparative analysis of reactive, proactive and hybrid routing protocols in MANET. Int. J. Electron. Comput. Sci. Eng. **1**(03), 1660–1667 (2012). (IJECSE, ISSN: 2277-1956)
8. Alam, M., Sher, M., Hussain, S.: An Integrated mobility model (IMM) for VANETs simulation and its impact. In: Proceedings of the IEEE International Conference on Emerging Technologies, pp. 452–456 (2009)
9. Ahmad, M., Khan, Z., Chen, Q., Najam-ul-Islam, M.: On the performance assessment of the OLSR protocol in mobile ad hoc networks using TCP and UDP. In: Hsu, C.-H., Xia, F., Liu, X., Wang, S. (eds.) IOV 2015. LNCS, vol. 9502, pp. 294–306. Springer, Heidelberg (2015). doi:10.1007/978-3-319-27293-1_26
10. Das, S., Raw, R.S., Das, I., Sahana, S., Purkayastha, B.S.: Effect of traffic density patterns on the performance of routing protocols for VANETs. In: 2015 International Conference on Computing, Communication and Automation (ICCCA), pp. 498–501. IEEE, May 2015
11. Chauhan, S., Tyagi, S.B.: Performance evaluation of reactive routing protocols in VANET (2014)
12. Perdana, D., Nanda, M., Ode, R., Sari, R.F.: Performance evaluation of PUMA routing protocol for Manhattan mobility model on vehicular ad-hoc network. In: 2015 22nd International Conference on Telecommunications (ICT), pp. 80–84. IEEE, April 2015
13. Alam, M., Sher, M., Hussain, S.A.: VANET mobility model entities and its impact In: 4th IEEE, International Conference on Emerging Technologies (ICET) NUST (2008)
14. Shaheen, A., Gaamel, A., Bahaj, A.: Comparison and analysis study between AODV and DSR routing protocols in VANET with IEEE 802.11b. J. Ubiquit. Sys. Pervasive Netw. **7**(1), 07–12 (2016)
15. Jabeen, Q., Khan, F., Khan, S., Jan, M.A.: Performance improvement in multihop wireless mobile adhoc networks. J. Appl. Environ. Biol. Sci. **6**, 82–92 (2016)
16. Khan, F., Khan, S., Khan, S.A.: Performance improvement in wireless sensor and actor networks based on actor repositioning. In: 2015 International Conference on Connected Vehicles and Expo (ICCVE), pp. 134–139. IEEE, October 2015
17. Khan, F., Nakagawa, K.: Performance improvement in cognitive radio sensor networks. Inst. Electron. Inf. Commun. Eng. (IEICE) **8**
18. Khan, S., Khan, F., Fahim Arif, Q., Jan, M.A., Khan, S.A.: Performance improvement in wireless sensor and actor networks. J. Appl. Environ. Biol. Sci. **6**, 191–200 (2016)
19. Khan, S., Khan, F., Khan, S.A.: Delay and throughput performance improvement in wireless sensor and actor networks. In: 2015 5th National Symposium on Information Technology: Towards New Smart World (NSITNSW), pp. 1–5. IEEE, February 2015

Secure and Safe Surveillance System Using Sensors Networks - Internet of Things

Fazlullah Khan[1]($\boxtimes$), Mukhtaj Khan[1], Zafar Iqbal[2], Izaz ur Rahman[1], and Muhammad Alam[3]

[1] Department of Computer Science, Abdul Wali Khan University Mardan, Mardan, Pakistan
{fazlullah,mukhtajkhan,izaz}@awkum.edu.pk
[2] Department of Computer Science, City University of Sciences and IT, Peshawar, Pakistan
zafariqbal@cusit.edu.pk
[3] Instituto de Telecomunicações, University of Aveiro, Aveiro, Portugal
alam@av.it.pt

Abstract. Sensor network is a network of autonomous devices that consist of sensors which are spatially distributed to sense the physical environment for certain parameters like temperature, humidity and pollution etc. There are various applications of sensor network, like volcanic eruption, inventory tracking system, military surveillance, homes and industrial automation and automobiles. Different sensors use for specific purpose such as temperature sensor, humidity sensor, light sensor, ultrasonic and multimedia sensor, and all these sensors are used for their own task. In this system, we use ultrasonic sensor for defense and security purpose. The ultrasonic sensor constantly transmits ultrasonic sound (Transmitter) which on striking with an obstacle bounces back and that bounced wave is also received by sensor (Receiver) and from this reflection the distance between sensor and obstacle is calculated. So when a person come close to dangerous area like electric field, river side and explosive material, the system will detect the person and will sound an alarm to inform the authorities. The proposed scheme is implemented and the generated results validates its functionalities.

Keywords: Internet of things · Security · Surveillance · Wireless sensor networks

1 Introduction

The "IoT" heralds the connection of a nearly countless number of devices to the internet thus promising accessibility, boundless scalability, amplified productivity and a surplus of additional paybacks [1]. Current real-world deployments of large-scale IoT systems are not limited to some well-bounded application domains. Sensor networks is one of the key network that will play a vital role to achieve the desired goals. Sensor networks uses in various areas because of

© ICST Institute for Computer Sciences, Social Informatics and Telecommunications Engineering 2017
J. Ferreira and M. Alam (Eds.): Future 5V 2016, LNICST 185, pp. 167–174, 2017.
DOI: 10.1007/978-3-319-51207-5_16

their unique characteristic that ranges from low level like mobile sensor (use in mobiles for call) to high level applications as nuclear plant monitoring. In sensor networks we deploy sensors in a field that is to be monitored for various parameter like temperature, humidity, pollution, light etc. the deployment criteria depends on application, it may be random or pre-planned [2]. The deployment in any hostile environment and in large geographical areas is usually random while in normal situation or limited areas we use pre-planned deployment technique.

Life of every human being is precious and the safety is a challenge for us. But the safety can be achieved by the use of various technological applications that do exist in this modern world. One of the techniques is through the deployment of sensor network. Therefore, we use ultrasonic sensor as a source of measuring the distance between human and the network. When a person approaches to define threshold the network will give a signal (alarm) to avoid the danger area. In this paper, the deployment of our network nodes are pre-planned as we have to monitor a specific area. The network can be deployed to any environment which can cause harm or threat to human life like electric field, border crossing, and huge water sides.

The reminder of this paper is structured as follows: the next section gives a short overview of relevant related work. Section 3 describes the proposed scheme. The implementation is described in Sect. 4. Section 5 gives an overview concerning the tests setup and the measurements results. In the last section, we draw the conclusions and some outlines for future work.

2   Related Work

We can find a number of related works focusing on the surveillance system using sensors networks such as a detail survey on multi-media can found it [3], a detail work on the energy efficient servilance system in [4]. The Australian Defense Force has IMAP and JMAP to perform planning prior to the deployment of forces, but there is a knowledge gap for on-ground forces during the execution of an operation [5]. Multi-agent based sensor systems can provide on-ground forces with a significant amount of real-time information that can be used to modify planning due to changed conditions. The issue with such sensor systems is the degree to which they are vulnerable to attack by opposing forces. This paper explores the types of attack that could be successful and proposes defense that could be put in place to circumvent or minimize the effect of an attack.

In [6], the authors state that it is practically impossible to construct a truly secure information system. Communications are secure if transmitted messages can be neither affected nor understood by an adversary, likewise, information operations are secure if information cannot be damaged, destroyed, or acquired by an adversary. They go on to define software challenges for a future combat system including (but not limited to) network security and accessibility; fault tolerance; and information analysis and summary of large data streams from the network. Further, author in [8] claim that most software is insecure. This could be because, as [9] have observed, security requirements are often omitted from requirements specifications altogether. This has been noted as being particularly problematic in other safety-critical domains such as automotive control software [10].

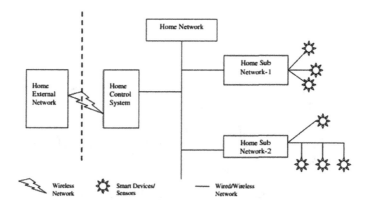

Fig. 1. Proposed system overview

In terms of the problem domain (military operations), wireless sensors of various types can be distributed on ground before a battle, whilst being connected to autonomous software agents in a multi-agent system to give an on-field tactical advantage, provided that the communications between the sensors cannot be subverted. A public key infrastructure is an obvious solution to the integrity problem, however issues of secure storage for the private key and over-the-air transmission of either public or private keys will still prove problematic. The issue of key management is perhaps further complicated by the ever-decreasing cost of the hardware required to conduct a brute-force attack [11].

3   Proposed Scheme

Our proposed model is focusing on the defense and security of individuals rather than a team. As stated, the basic application is to comfort and ease in life of general public as it can be installed in almost all places with low-cost and operation facilities. The previous work in this field were about high level security i.e. on state level surveillance but our model will provide the security in our routine work. In this model, we define an threshold and permitted area, where if someone tries to get into that particular area the alarm system will invoke the security officials as well as the individuals living or staying in that particular area on that particular time.

The proposed algorithm for the scheme is working on the principle that when the sound signal is generated by ultrasonic sensor and it echo back receive by the receiver of ultrasonic sensor and send to micro controller, the micro controller calculate the time. The time at which sound is produced and which it is received are also recorded. The distance is calculated on the bases of this time. The formula for the distance is

$$D = \frac{\frac{time \, in \, \mu \, sec}{73.746}}{2} \, inches \tag{1}$$

Fig. 2. An HC-SR04 type ultrasonic sensor

Using the above distance formula, we calculate threshold "t" value, where as for t < 24 in., it will display warning massage to the base station and will sound an alarm. An overview of the proposed system is presented in the Fig. 1. The figure shows that sensor network is integrated with the home sub-network and that is further connected with the home networks. Inside home, wireless and wireless networks can be used to connect with the external networks. For instance, if the users want to save the events in a server they can use the in-house networks to store the activities in the server.

4   Implementation

This section overviews the variety of hardware and software used to implemented the proposed scheme and generating results.

4.1   Ultrasonic Sensor

This type of sensors generates high frequency sound waves and evaluates the echo which is received back by the sensor. The frequency of sound wave is about 20 KHz or above. The time interval between sending the signal and receiving the echo determines the distance of object. Figure 2 depicts the sensor that has been used in the tests.

Fig. 3. Micro-controller chip used.

Fig. 4. Proposed system overview

4.2 Micro Controller

A micro controller is a small computer as it has a single IC. The micro controller has its own processor core, memory. The function of micro controller is to process the data (Fig. 3).

4.3 Bread Board and Jumper Wires

A bread board is used for making an experimental model of an electric circuit. As micro controller can support only a few devices therefore we use bread board

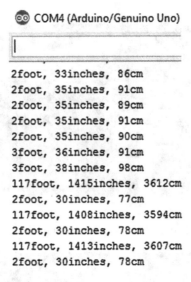

Fig. 5. Output in normal condition

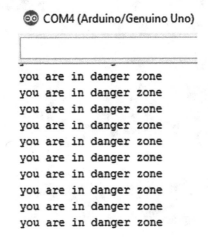

Fig. 6. Output at the time of intruder detection

to connect multiple devices with the micro controller through jumper wires for initial test at laboratory level.

4.4   Software

We have used the open source Arduino Integrated Development Environment (IDE). The open-source Arduino Software (IDE) makes it easy to write code and upload it to the board. It can run on Windows, Mac OS X, and Linux.

5   Installation of Tools and Results

As stated earlier, the different devices are interconnected with micro controller through bread board. The connection was provided by jumper wires. The micro controller is further connected to laptop to display the result. The system is shown in Fig. 4.

When the network is in off state, it does not sense any movement and does not generate any data or information. Initially, the system is tested for normal condition and the situation when there is someone near the in vicinity. In normal condition, the system will sense the data through sensors; the micro controller will process it and will pass the output to the base station. The micro controller processes the data (calculate the distance) and take a decision on the calculated data. In normal condition when the threshold is not reach network will not take any action. As depicted in Fig. 5, the system shows the result of detection at various distances from the system.

When the defined threshold is reached i.e. the distance between network and intruder is less than defined threshold the system will sound an alarm and lights will start blinking and the message shall be displayed as "danger area" as depicted in Fig. 6. It should be noted that based on the scenario and application, we can change the defined threshold.

6   Conclusions and Future Work

In this paper, we presented a scheme based on sensor networks for surveillance system. In a nut-shell, the presented system is based on sensor network which use ultrasonic sensor and it take decisions based on sensed data at define threshold. The proposed system is implemented and validated. In future, we are committed to use the ultrasonic sensor in other applications as auto door's opening, and home automation.

References

1. Alam, M., Ferreira, J., Fonseca, J.: Introduction to intelligent transportation systems. In: Alam, M., Ferreira, J., Fonseca, J. (eds.) Intelligent Transportation Systems. SSDC, vol. 52, pp. 1–17. Springer, Heidelberg (2016). doi:10.1007/978-3-319-28183-4_1
2. Alam, M., Rodriguez, J.: A dual head clustering mechanism for energy efficient WSN. In: Proceedings of the 2nd International Conference on Mobile Lightweight Wireless Systems - MOBILIGHT, Barcelona, Spain, May 2010
3. Cucchiara, R.: Multimedia surveillance systems. In: Proceedings of the Third ACM International Workshop on Video Surveillance and Sensor Networks, pp. 3–10. ACM (2005)
4. Tian, H., Krishnamurthy, S., Stankovic, J.A., Abdelzaher, T., Luo, L., Stoleru, R., Yan, T., Gu, L., Hui, J., Krogh, B.: Energy-efficient surveillance system using wireless sensor networks. In: Proceedings of the 2nd International Conference on Mobile Systems, Applications, and Services, pp. 270–283. ACM (2004)

5. Johnstone, M.N., Thompson, R.: Security aspects of military sensor-based defence systems. In: 2013 12th IEEE International Conference on Trust, Security and Privacy in Computing and Communications, pp. 302–309. IEEE (2013)

6. Wysopal, C., Nelson, L., Dai Zovi, D., Dustin, E.: The Art of Software Security Testing. Addison Wesley, Upper Saddle River (2007)

7. RFC 2828: Internet security glossary. Internet Engineering Task Force (2016). http://www.ietf.org/rfc/rfc2828.txt. Accessed 22 June 2016

8. Lozano, A., Jindal, N.: Transmit diversity vs. spatial multiplexing in modern MIMO systems. IEEE Trans. Wirel. Commun. 9(1), 186–197 (2010)

9. Puthal, D., Nepal, S., Ranjan, R., Chen, J.: A dynamic key length based approach for real-time security verification of big sensing data stream. In: Wang, J., Cellary, W., Wang, D., Wang, H., Chen, S.-C., Li, T., Zhang, Y. (eds.) WISE 2015. LNCS, vol. 9419, pp. 93–108. Springer, Heidelberg (2015). doi:10.1007/978-3-319-26187-4_7

10. Jan, M.A., Nanda, P., He, X., Liu, R.P.: A lightweight mutual authentication scheme for IoT objects. IEEE Trans. Dependable Secure Comput. (TDSC), pp. 670–676 (2016)

11. Puthal, D., Nepal, S., Ranjan, R., Chen, J.: A dynamic prime number based efficient security mechanism for big sensing data streams. J. Comput. Syst. Sci. 83(1), 22–42 (2016)

Challenges and Opportunities in Big Data and Cloud Computing

Hassan Sohail, Zeenia Zameer, Hafiz Farhan Ahmed, Usama Iqbal,
and Pir Amad Ali Shah[(✉)]

Computer Science Department, University of South Asia, Lahore, Pakistan
hassansohail@live.com, zeenia.zameer@yahoo.com,
farhanahmad1525@gmail.com, uiqbal970@gmail.com,
amad.ali@usa.edu.pk

Abstract. In this paper we have discussed the approaches and complexity of big data analytics issues in the perspective of cloud computing. Big data and cloud computing are novel methods of distributing computational resources. Currently big data and cloud computing has engrossed romantic behavior, which needs handling huge quantity of data quickly and securely. In the modern era the size of data is dramatically increasing due to cloud computing technology, which poses various challenges with the variety, security, and size of data. Cloud computing is the newest and the main kind of distributed computing systems and it wraps most of their features. It has already been broadly employed for its huge merits and its capability to handle large amount of data like workflows and big data applications. We have identified some gaps in technology, its challenges, limitation, and applications of big data and cloud computing from theoretical perspective. Moreover, this paper give recommendations to the researchers on future direction sand suggests different solutions to the challenges and limitations.

Keywords: Big data · Cloud computing · Cloud data · Security

1 Introduction

The advancement in science and technology has made the globe as a small town and as a result a huge amount of data is produced and stored daily. Extracting precise and relevant information from this data is very challenging and helpful in business competition. Various data mining solutions that extract structured and unstructured data is main key for organizations to gain insights from company private data as well as huge amounts of publicaly available data. The capability of validating customer private information such as product preferences, likes and dislikes etc. with the data obtained from tweets, blogs, feedback, product evaluations, and information available on social networks opens a broad variety of potentials for organizations to know the requirements of their clientele, forecast their needs, demands, and efficiently utilize the resources. This standard is being popularly known as Big Data. In other words, the big data is described as a dataset whose volume is afar the processing capability of classical databases or systems. Four essentials are highlighted in the description of big data that

© ICST Institute for Computer Sciences, Social Informatics and Telecommunications Engineering 2017
J. Ferreira and M. Alam (Eds.): Future 5V 2016, LNICST 185, pp. 175–181, 2017.
DOI: 10.1007/978-3-319-51207-5_17

are capture, store, manage, and analyses [2]. The focal point of the four essentials is the final phase, the big data analytics that is repeatedly pulling out information from a huge quantity of data. It can be seen as the drawing out or handing out of the enormous data and helpful facts can be regained from the large dataset [3]. The conventional technique for analyzing data is laid on the arithmetical models of the problems first and then looks if data vigorous the models. With the increase of the diversity of sensible data, these arithmetical models might be unproductive in resolving problems. The pattern must move from the model-driven to the data-driven techniques. The data-driven technique spotting on forecasting what is happening and also ponders on what is occurring right now and getting prepared for the potential events.

Cloud computing has been transfiguring the Information Technology industry by giving litheness to the way Information Technology is utilized, allowing organizations to pay only for the services and resources they employ. In an attempt to lessen IT assets and running cost, from small to large size are using Clouds to offer the wherewithal needed to run their applications. Clouds differ considerably in their precise technologies and implementation, but frequently give infrastructure, platform, and software resources as services. The most frequently stated benefits of Clouds include offering resources in a pay-as-you-go fashion, enhanced accessibility and flexibility, and fee decrease. Clouds can avoid organizations from costs for preserving peak-provisioned IT assets that they are improbable to utilize the majority of the time. Even as at first glimpse the price suggestion of Clouds as a podium to perform analytics is strong, there are numerous issues that require trouncing to craft Clouds a perfect platform for scalable analytics. In cloud computation, usually many computers are being used to lodge all the users' requirements on instantaneous base. So an enormous facts and/or figures have to be transmitted from one place to another place for implementation of some programs based on the necessities of memory, processor, disk space etc. especially in big data Cloud computing network have some attractive manners like servers are clustered as sub-network within the network. The arrangement can be shaped to specific arrangement techniques. Cloud computing usually has two kinds of connection, i.e. direct and indirect. Peer-to-peer data movement is handling by direct connections while observing system holds all the situations in the network using indirect connections. Fetching huge amount of data is hard as there are many barriers to conquer. The first barrier is handling the gigantic quantity of data rapidly. The size of data influences the performance of cloud network. The big data analytics also experiences from problems where the huge amount of data will be executed in a short time with a sensibly good performance.

The rest of this paper is organized as follows. In Sect. 2, challenges and issues is presented followed by the discussion in Sect. 3. The paper in concluded and future research directions and gaps are discussed in Sect. 4.

2 Challenges and Opportunities

With any kind of advancement in technology, cloud computing ought to be thoroughly assessed prior to its extensive adoption. A small number of researches have scientifically considered cloud computing affect on information technology by classifying it as challenges of cloud computing and its opportunities. In this paper we assess the opportunities and challenges from the following perspective; with opportunity and challenges [6–8].

1. Management as a challenge in Big Data and Cloud Computing
    a. Less expenses on Information Technology infrastructure
        i. Lack of trust by health care professionals
    b. Computing resources available on demand
        i. Organizational inertia
    c. Payment of use on a short-term basis as needed
        i. Loss of governance
2. Technology as a challenge in Big Data and Cloud Computing
    a. Reduction of IT maintenance burdens
        i. Uncertain provider's compliance
    b. Scalability and flexibility of infrastructure
        i. Resource exhaustion issues
        ii. Unpredictable performance
    c. Benefits for green computation
        i. Secure Data
        ii. Restricted access on data transfer
        iii. Faults in hefty distributed cloud computing
3. Security as a challenge in Big Data and Cloud Computing
    a. No resource constraints on protecting data
        i. Non-centralized collapse
    b. Secure data by placing its various copies at different locations
        i. Public administration problems
    c. Vigorously leveled protective assets intensification pliability
        i. Meager key encryption
4. Legal as a challenge in Big Data and Cloud Computing
    a. Supplier's promises to guard consumer's information/privacy
        i. Privilege abuse
    b. Mature strategies/technologies for enabling the erection of reliance policies by nonprofit groups
        i. Data jurisdiction issues
    c. Development of rules by government regarding information/privacy security
        i. Privacy issues
5. Scalability as a challenge in Big Data and Cloud Computing
    a. Distributed data storage systems
        i. Relational Database Management Systems are not supported by cloud technologies
6. Availability as a challenge in Big Data and Cloud Computing
    a. Data is accessible from everywhere
        i. deliver high-quality services
        ii. data integrity and security

7. Transformation as a challenge in Big Data and Cloud Computing
   a. Structured and Unstructured data
      i. In unstructured data, data need to saved in a distributed databases prior to processing
8. Heterogeneity is a challenge in Big Data and Cloud Computing
   a. Data from multiple sources
      i. users can save data three formats, i.e. structured format, semi-structured format, or unstructured format
9. Privacy as a challenge in Big Data and Cloud Computing
   a. Personal details are one click away
      i. Personal details are exposed to scrutiny,
      ii. Profiling, stealing, and loss of control issues
10. Governance is a Challenge in Big Data and Cloud Computing
    a. Use of standards
       i. Applications that use huge amounts of data obtained from outside sources.

3 Discussion

The key limitations and challenges in big data and cloud computing are the amount of data and the storage capacity, accessing and fetching the data. Cloud computing usually uses direct and indirect connections. Small amount of data on direct connections have no issues however when the amount of data increases then fetching large dataset has different hurdles to overcome. The first hurdle is the amount of data which ultimately affect processing on data. The big data analytics experiences from problems where the huge amount of data will be executed in a short time with a sensibly good performance. Another hurdle is the frequency of changes in the data content. As the data is constantly growing so efficient and intelligent algorithms are need to be modified. Likewise, another hurdle is the diversity of data, i.e. various types' data coming from various sources. Sometimes various types of formless data require to be pre-processed to semi-formed and/or well formed data before final processing. In some situations, several goals need to be attained concurrently in large datasets. Mainstream conventional techniques can only produce good result with constant and various algorithms, and need to do a sequence of disconnect runs to assure different goals [7, 8]. With the advancements in various communication fields such as intelligent transportation systems [9], Internet of Things (IOT), Software defined networks, there is requirement for more study to break the multi-goals problems with less constraint.

The concept of cloud computing shift the computing power and data storage away from computer into the cloud, here are the potential benefits of cloud computing using swarm intelligence for big data. It helps to overcome processing limitations of normal systems with high amount of data storage, and power consumption. It also increases security level for data through a centralized monitoring and maintenance of software. Similarly, the power of the cloud computing can be seen in E-commerce and E-marketing where huge amount of data is produced. For example, use of the cloud technology in business allows us to efficiently utilize resources which reduce business cost. Thus it will charm enterprise's kernel competitive control and finally complete goods and services trading.

Further, we give detail discussion of the challenges and opportunity discussed in previous section, i.e. management, technology, security, legality, scalability, availability, transformation, heterogeneity, privacy, and governance.

1. **Management**

   The principle advantage of cloud computing is its low cost due to its quick deployment and maintenance. This ability shows that due to change in demand organizations are not required to change infrastructures. Similarly a cultural confrontation; to share information and changing conventional ways of working is a general administration challenge to accept cloud technology.

2. **Technology**

   Smaller organizations typically do not have internal IT staff. Hence, removing the new information technology infrastructure and maintenance burdens can eliminate lots of barriers. Similarly, cloud computing has advantages for so-called *green computing*—the more efficient use of computer resources to help the environment and promote energy saving. However, the main issues are resource exhaustion, unpredictability of performance, data lock-in, data transfer bottlenecks, and bugs in large-scale distributed cloud systems.

3. **Security**

   Possibly the toughest antagonism to cloud technology implementation in an organization is data security. Most cloud suppliers put users' information in various locations which makes profile redundant and less secure. These includes hacker attacks, network breaks, natural disasters, separation failure, public management interface, poor encryption key management, and privilege abuse.

4. **Legality**

   Data and privacy protection are necessary for gaining consumer faith required for cloud technology to utilize its complete prospective. If the supplier implements enhanced and efficient rules and practice, consumer will easily evaluate the associated risks they face. However, the use of cloud technology may pose various legal issues like contract law, intellectual property rights, data jurisdiction, and privacy.

5. **Scalability**

   The capability of memory to grip growing sums of information in a suitable way. Scalable dispersed data storage systems is a vital element of cloud infrastructures. However, due to the lack of cloud computing features to support RDBMS associated with enterprise solutions has made RDBMS less attractive for the deployment of large-scale applications in the cloud.

6. **Availability**

   It is the resources of a system available to the clients on demand by an authoritative person. In a cloud computing, a major problem regarding cloud service providers is the accessibility of data saved in the cloud. However, the increase in number of cloud users may cause quality of services, and most important issue is the private and secret information on cloud has security threats.

7. **Transformation**

Altering the data into an appropriate format for investigation is a hindrance in the implementation of big data. Due to different types of data formats, big data can be changed into an analysis workflow is different for structured and unstructured data.

8. **Heterogeneity**

Data from multiple sources are generally of different types and representation forms and significantly interconnected; they have incompatible formats and are inconsistently represented, i.e. users can store data in structured, semi-structured, or unstructured format. However, unstructured data are inappropriate because they have a complex format that is difficult to represent in rows and columns.

9. **Privacy**

Privacy apprehensions carry on hampering those clients who outsource personal data into cloud. This apprehension has turned into severe issues with the improvement of big data mining and analytics, which need personal information to produce relevant results, such as personalized and location-based service. However, if Information of an individual is leaked which will gives rise to concerns on profiling, stealing, and loss of control.

10. **Governance**

The exercise of control and authority over data related rules of law, transparency, and accountabilities of individuals and information systems to achieve business objectives. However, policies, principles, and frameworks that strike stability between risk and value in the face of increasing data size and deliver better and faster data management technology can create huge challenges.

4   Conclusion

In this paper we have mentioned a few important challenges and issues in big data and cloud computing. We discussed cloud computing will grow and with the age of big data, and have elaborated some of the key challenges existing in the field of cloud computing. With the existing tool and techniques it is not sufficient to adhere all the challenges relating to big volume of data. It is not feasible to provide better data quality with the existing technology and again privacy is big problem with cloud data. Processing streaming data need some novel algorithms and some efficient tools. Again from privacy point of view hashing is not possible for data with volume of data. Confidentiality challenge for data can be addressed by developing any novel algorithms for key management and key exchange.

References

1. Wang, Z., Chu, Y., Tan, K.-L., Agrawal, D., Abbadi, A.E., Xu, X.: Scalable data cube analysis over big data. arXiv preprint arXiv:1311.5663 (2013)
2. Schroeck, M., Shockley, R., Smart, J., Romero-Morales, D., Tufano, P.: Analytics: The Real-World Use of Big Data. IBM Global Business Services, Armonk (2012)

3. Akerkar, R.: Big Data Computing. CRC Press, Boca Raton (2013)
4. Majhi, S.K., Shial, G.: Challenges in big data cloud computing and future research prospects: a review. Smart Comput. Rev. **5**(4) (2015)
5. Assunção, M.D., Calheiros, R.N., Bianchi, S., Netto, M.A.S., Buyyab, R.: Big data computing and clouds: trends and future directions. J. Parallel Distrib. Comput. **79–80**, 3–15 (2015)
6. Kuo, A.M.-H.: Opportunities and challenges of cloud computing to improve health care services. JMIR **13**(3) (2011)
7. http://sameekhan.org/pub/H_K_2015_IS.pdf
8. Targio, H.I.A., Yaqoob, I., Anuar, N.B., Mokhtar, S., Gani, A., Khan, S.: The rise of big data on cloud computing: review and open research issues. Inf. Syst. **47**, 98–115 (2015)
9. Alam, M., Ferreira, J., Fonseca, J.: Introduction to intelligent transportation systems. In: Alam, M., Ferreira, J., Fonseca, J. (eds.) Intelligent Transportation Systems. SSDC, vol. 52, pp. 1–17. Springer, Heidelberg (2016). doi:10.1007/978-3-319-28183-4_1

Applications and Challenges Faced by Internet of Things - A Survey

Pir Amad Ali Shah[1], Masood Habib[2(✉)], Taimur Sajjad[2],
Muhammad Umar[2], and Muhammad Babar[3(✉)]

[1] Department of Computer Science, University of South Asia, Lahore, Pakistan
amad.ali@usa.edu.pk
[2] Department of Computer Science,
Comsats Institute of IT Sahiwal, Sahiwal, Pakistan
{masoodhabib,taimursajjad,mumar}@ciitsahiwal.edu.pk
[3] National University of Sciences and Technology, Islamabad, Pakistan
babarkhan666@hotmail.com

Abstract. The Internet of Things (IoT) is a concept which expands the extent of internet by integrating a physical object to discover them into contributing bodies. This novel idea allows a physical gadget to embody itself in the digital world. There are a bunch of conjectures and opportunistic future of the IoT devices. However, most of them are vendor specific and requires a cohesive standard, which delivers their flawless assimilation and interoperable operations. Another key issue is the need of highly secure features in these devices and their equivalent products. Majority of these devices are resource constrained and not able to sustain computationally intricate and resource overwhelming secure algorithms. In this paper, we present a survey of various applications which have been made possible by IoT. Furthermore, the challenges faced by these networks.

Keywords: Internet of things · IETF · CoRE · CoAP

1 Introduction

Twenty first century has revolutionized the world of technology. Size of internet has been increasing rapidly with integration of miniaturized embedded devices into the internet world. Automation systems, personal gadgets, smart grid, cell phones and many other devices collaborate with each other and share valuable information about physical world. Internet is moving from traditional workstation and laptops to small embedded devices. We are moving from internet to Internet of Things (IoT) [1] by incorporating a sheer number of physical devices into internet. These objects contain miniature sensor nodes at their core which inherits all the limitations of Wireless Sensor Networks. IoT extends internet beyond personal computers, work stations to the world of physical objects. A broad range of appliances are now connecting to internet and provides valuable information. In internet, humans are the main source of generating information ranging from sending emails, capturing videos to messaging and browsing are some to mention. However, in IoT of the future, there will be millions and trillions of smart objects which will collect information, process it and communicate it. IoT

© ICST Institute for Computer Sciences, Social Informatics and Telecommunications Engineering 2017
J. Ferreira and M. Alam (Eds.): Future 5V 2016, LNICST 185, pp. 182–188, 2017.
DOI: 10.1007/978-3-319-51207-5_18

relies of a set of distinct technologies which collaborate with each other. The major technologies behind this vision of IoT are Identification, sensing, embedded processing and communication are some to mention [2]. Radio Frequency Identification (RFID) tags are attached to physical world objects which contain data about those objects. These small tags are not capable to sense the environment but have the ability to collect data about a product. Internet of Things would not have been possible without them as they provide each object a unique identification to be recognized on the internet. On the other hand, sensor networks are capable to sense the environment based on unique identification provided by RFID tags and can also monitor their location, energy and other parameters. Once data is being sensed, partial processing take place at each object which is further transmitted for various operations to extract valuable information from it.

The Internet Engineering Task Force (IETF) shaped a working group called Constrained RESTful Environment Group (CoRE) group. This group was given the job to define a method to use a large number of tiny, resource limited, low-power devices, that can exchange information over lossy- networks. This group described a set of regulations that is termed as Constrained Application Protocol (CoAP) [3]. CoAP is an application-layer protocol that is created to permit information exchange between resource-limited gadgets over resource limited networks [4]. Resource limited devices are tiny devices which have low processing power, memory, and speed. These devices mostly manufactured with 8-bit microcontrollers. CoAP protocol runs over UDP and cannot use TCP. IPv6 over Low-Power Wireless Personal Area Network (6LoWPAN) is an example of such a constrained network configuration setup [5]. CoAP has similar to HTTP-like request and response paradigm where devices can interact by sending a request and receiving a response. CoAP is very similar to HTTP; it is evident that it has been intended for easy web integration. CoAP does not replace HTTP; instead, it implements a small subset of widely accepted and implemented HTTP practices and optimizes them for M2 M message exchange. Think of CoAP as a method to access and invoke REST fu l services exposed by "Things" over a network [6]. Some good survey and research on intelligent transportation and its application in Internet of Things can be studied [12–14].

In this paper, we discuss the applications of IoT and the wide range of challenges faced by these networks. The applications of IoT are restricted by various challenges faced by these networks at various layers. The rest of the paper is organized as follows. In Sect. 2, potential applications are discussed. In Sect. 3, various challenges faced by these networks are discussed. Finally, the paper is concluded and future research directions are provided in Sect. 4.

2   Applications of Inter of Things

The IoT allows us to use technology to improve our reassure, efficiency utilizes our energies, and easily performs the tasks that utilize our home and work life and give us greater control over our lives. Here, we discuss various applications of an IoT.

2.1   Connected Home

A connected home can mean dissimilar things to different people, but it is basically a home with one or more gadgets linked together so that the homeowner can organize, modify and check their environment. If the IoT is basically helping our lives comfortable and easier and more linked, then the connotations for a truly Connected Home are game-changing.

2.2   Wearable

This technology covers a verity of devices that monitors, record, and give response on you/your environment. In other words, wearable are divided in two categories:

2.2.1   Fitness and Environment
Fitness ribbons and wristwatches are capable to monitor and send data based on your daily activities such step counting, heart rate and temperature.

2.2.2   Health
These wearable devices can monitor vital health factors such as O_2 saturation, heart beat etc, and can transmit any information outside of a planned range to the patient and to his doctor.

2.3   Industrial IoT

The IoT has thoughtful solution to automate an industry with wireless and infrastructure-less connectivity using sensor networks, M2 M communications, and conventional industrial automation can be made efficient and more effective.

2.4   Smart Grid

A smart grid is collection of internet capable devices that could measures power/energy, water or natural gas utilization of a town/building. With smart grid we can save labor cost as well as actual and accurate information and demands of the users.

2.5   Transportation

CoAP protocol is used for tracking the vehicle by fetching the GPS coordinates of the vehicle position at a specific point of time. It monitors the speed of the vehicle by fetching the reading of the accelerometer of the vehicle. A simple symmetric handshaking for various states of the vehicle (Fast moving, slow moving and Rest) is investigated. The overhead incurred during the communication and handshaking is quiet low which suits the requirement of energy constrained devices.

These are just a subset of applications. There are many other applications of IoT. The scope and nature of IoT provides a wide range of opportunities for various applications. Currently, a wide range of research is being conducted to investigate the applications of CoAP and various other IoT protocols for physical objects of daily life.

3 Challenges Faced by Inter of Things

Internet of Things consists of a bunch of physical devices connected with each other. The devices themselves are resource-rich; however, they will not be able to communicate with each other in absence of sensor nodes. The presence of sensor nodes at the core of each physical device makes the device intelligent and enables it to identity itself in the digital world. These sensor nodes are resource-constrained and as a result classify the device as resource-constrained as well [7]. Resource-constrained devices vary from one another in term of space code, RAM and other specifications which affects their capabilities to support HTTP protocol. Resource-constrained devices having 10 KBytes of RAM and about 100 Kbytes of ROM are not capable to support HTTP (Class 1 devices) while those having 50 Kbytes of RAM and around 250 Kbytes of ROM support HTTP (Class 2 devices) [9]. However, HTTP requires considerable amount of code space and ROM along with high energy in processing, so Class 1 devices refrain from adapting HTTP. As a result, extremely lightweight protocols such as CoAP need to be developed to make them feasible for IoT. The protocols need to adjust the battery power of each object so that they can operate for months and years for as little as 1 W.

Another challenging issue for IoT is interoperability [8]. As IoT incorporate a series of devices, hence, interoperability between various devices is a serious issue. Most of these objects have their own underlying hardware and software platforms and as a result, they are not able to communicate with each other. As a result, a common and unified standard for various technologies is required. The use of such a standard will provide seaming-less operations.

The devices require a scalable application layer for interoperable communication. Moreover, a common programming model is required, which will enable programmers to focus only on application development rather than the hassle of worrying about underlying platform architecture. In [15], the authors proposed an innovative solution to cope with these challenges by curbing the installation of application code on the embedded systems. Rather, they suggest that application code should run on the cloud and only firmware and network stack will be nested in the core of each embedded device. Running applications on cloud will serve two major purposes: ample memory space availability on the nodes and most importantly, developer will not have to worry about the hardware architecture [8]. The latter will help to develop applications on cloud which will enhances communication between heterogeneous nodes irrespective of any programming language. RESTful operations will be performed on the nodes to communicate with the hardware and perform various operations. Cloud operation will enhances communication between nodes from different manufacturers and will provide an interoperable communication between them. Now a NetDuino board will not require a custom protocol to communicate with TMote or Berkeley mote as everything is running on the code. Only Firmware and RESTful operations (PUT, DELETE, GET, and POST) will all that be implemented on the node. Application code is shifted to cloud. In-network data processing consume considerable amount of a node resources, these operation will also need to be shifted to a powerful devices in order to ease the burden on these nodes [2].

The Quality of Service (QoS) provisioning in an IoT framework is another challenging issue which needs to be addressed. To provide QoS, two parameters are of high importance: Reliability and Timely delivery of data. Reliability is provided by transmitting CON messages (message type in CoAP) which need to be acknowledged. When a sender transmits a CON message to the server for resource observation (resources such as temperature etc. resides on a server), it waits for an acknowledgement by using Stop-and-Wait retransmission algorithm. In Resource observation, timeliness is maintained by using "Observe" option. This option enables the subscriber/listener to sequence the notification.

In resource observation, an observer registers itself with a resource residing on a server [10]. The subject (server) notify each observer when the state of the resource changes. This reduces the number of transmission flowing in the network which in turn improves the efficiency, reliability, energy consumption, bandwidth utilization and other QoS metrics of the network. Resource observation provides reliability by exchanging CON messages which need to be acknowledged. As far as timeline requirement is concerned, the Observe option helps the observers in sequencing the resources. Though, this option helps the subscribers/observers to check the validity of the notifications. However, it does not guarantee timely delivery of notification (carrying resources) to the observers. This will have severe consequences in real time application where a minor delay in notification will make it useless [11].

The presence of diverse range of devices at the core of IoT poses various security threats. Integrating everyday objects into the internet require various communication models. This requirement is likely to add some very ingenious and innovative malicious models [16]. It is of utmost importance that such models should be prevented or at least mitigating options should be in place to tackle their undesirable effects. To develop a secure solution in the internet of things context is much more difficult due to varying and unpredictable nature of objects, many of whom are to be connected for the first time in the internet. It is very important to understand the characteristics and features of things and underlying embedded technologies to combat various malicious models. Existing security and lightweight cryptographic algorithms are to be assessed and adapted in the internet of things environment. However, such profiling of these protocols and algorithms might not necessarily comply with their domain of applications and might results in undesirable outcomes. Any protocol or algorithm has their intended domain of applications and specification. Modification of protocol features might deviate from its original use of intend as many internet-based protocols were not designed for internet of things objects. Recent work can found in [10].

Heterogeneity plays a vital role in infrastructure protection. Highly constrained sensor nodes scattered in a battle field require a robust communication channel to communicate with cellular and wireless devices like Smartphone. Cryptographic algorithms are required to secure communication between these entities. However, due to battery power nature of these devices, the algorithms need to be computationally simple and fast efficient. AES algorithm might suit a small subset of IoT devices; however, they might not be suitable for extremely constrained RFID tags. Symmetric algorithms are the best options rather than asymmetric algorithms as they are computationally simple and suit these tags etc [9, 10]. IoT devices need to use the existing internet standards to communicate with each other. However, all of them are not resource oriented. Hence existing security protocols need to be adapted and modified. In short, the challenges faced by IoT are summarized as follows:

- Sensing a complex environment: Innovative ways to sense and deliver information from the physical world to the cloud
- Connectivity: Variety of wired and wireless connectivity standards are required to enable different applications needs.
- Power is critical: Many IoT applications need to run for years over batteries and reduce the overall energy consumption.
- Security is Vita l: Protecting user's privacy and manufacturers IP, detecting and blocking malicious activities.
- IoT is complex: IoT application development needs to be easy for all developers, not just to exports.
- Cloud is important: IoT applications require end-to-end solutions including cloud services.

4 Conclusion

Internet of things incorporates a wide range of devices. The presence of miniature sensor nodes at the core of each device provides seamless and interoperable communication. Although, a wide range of applications exist, however, communication is still at risk in these applications. These networks face various challenges which need to be addressed in order to broaden the scope of IoT. These networks have the potential to enable communication between devices, which were not previously connected with the internet.

References

1. Atzori, L., Iera, A., Morabito, G.: The internet of things: a survey. Comput. Netw. **54**(15), 2787–2805 (2010)
2. Usman, M.J., Zhang, X., Chiroma, H., Abubakar, A., Gital, A.Y.: A framework for realizing universal standardization for internet of things. J. Ind. Intell. Inf. **2** (2014)
3. Understanding Constrained Application Protocol Using CoAPSharp Library (2014). www.coapsharp.com
4. Jan, M.A., Nanda, P., He, X.: Energy evaluation model for an improved centralized clustering hierarchical algorithm in WSN. In: Tsaoussidis, V., Kassler, Andreas, J., Koucheryavy, Y., Mellouk, A. (eds.) WWIC 2013. LNCS, vol. 7889, pp. 154–167. Springer, Heidelberg (2013). doi:10.1007/978-3-642-38401-1_12
5. Jabeen, Q., Khan, F., Khan, S., Jan, M.A.: Performance improvement in multihop wireless mobile adhoc networks. J. Appl. Environ. Biol. Sci. (JAEBS) Int. J. Eng. Trends Appl. (IJETA) **3**(2) (2016)

6. Shelby, Z., Bormann, C.: 6LoWPAN: The Wireless Embedded Internet, vol. 43. Wiley, Hoboken (2011)
7. Kovatsch, M.: Firm firmware and apps for the internet of things. In: Proceedings of the 2nd Workshop on Software Engineering for Sensor Network Applications, pp. 61–62. ACM (2011)
8. Puthal, D., Nepal, S., Ranjan, R., Chen, J.: DPBSV–an efficient and secure scheme for big sensing data stream. In: 2015 IEEE on Trustcom/BigDataSE/ISPA, vol. 1, pp. 246–253. IEEE, August 2015
9. Puthal, D., Nepal, S., Ranjan, R., Chen, J.: A dynamic key length based approach for real-time security verification of big sensing data stream. In: Wang, J., Cellary, W., Wang, D., Wang, H., Chen, S.-C., Li, T., Zhang, Y. (eds.) WISE 2015. LNCS, vol. 9419, pp. 93–108. Springer, Heidelberg (2015). doi:10.1007/978-3-319-26187-4_7
10. Jan, M.A., Nanda, P., He, X., Liu, R.P.: A lightweight mutual authentication scheme for IoT objects. IEEE Trans. Dependable Secure Comput. (TDSC) (submitted, 2016)
11. Puthal, D., Nepal, S., Ranjan, R., Chen, J.: A dynamic prime number based efficient security mechanism for big sensing data streams. J. Comput. Syst. Sci. **83**, 22–42 (2016)
12. Jan, M.A., Nanda, P., Usman, M., He, X.: PAWN: a payload-based mutual authentication scheme for wireless sensor networks. 15th Int. J. Eng. Trends Appl. (IJETA) **3**(2) (2016)
13. Alam, M., Ferreira, J., Fonseca, J.: Introduction to intelligent transportation systems. In: Alam, M., Ferreira, J., Fonseca, J. (eds.) Intelligent Transportation Systems. SSDC, vol. 52, pp. 1–17. Springer, Heidelberg (2016). doi:10.1007/978-3-319-28183-4_1
14. Alam, M., Albano, M., Radwan, A., Rodriguez, J.: CANDi for energy saving and facilitating short-range. Trans. Emerg. Telecommun. Technol. **26**, 861–875 (2013). doi:10.1002/ett. 2763. ISSN 2161-3915
15. Alam, M., Trapps, P., Mumtaz, S., Rodriguez, J.: Context-aware cooperative testbed for energy analysis in beyond 4G networks. Telecommun. Syst. J. **62**, 1–20 (2016). doi:10.1007/ s11235-016-0171-5. Online ISSN 1572-9451
16. Explore What's Possible with the Internet of Things. http://www.silabs.com/solutions/iot. html

Cloud Computing Development Life Cycle Model (CCDLC)

Muhammad Babar$^{(\boxtimes)}$, Ata ur Rahman, and Fahim Arif

National University of Sciences and Technology, Islamabad, Pakistan
babarkhan66@gmail.com, atabcs_87@yahoo.com,
fahim@mcs.edu.pk

Abstract. Cloud Computing is getting reputation as the standard approach for designing and organizing software applications over the internet, especially for distributed and e-commerce applications. In recent times, Cloud Computing has emerged as a new opportunity that how software and other resources can be provided to the consumers as a service. However, applying the existing traditional software engineering life cycle models to cloud computing, we identify some inadequacies like they do not concentrate on engineering activities, they lack the fundamental description of cloud services using traditional requirement engineering process, they do not deal with proper modeling, and they suffer from good development and testing processes. We propose a cloud computing life cycle model (CCDLC) for development of cloud to deal with the mentioned deficiencies faced in the existing traditional life cycle models.

Keywords: Cloud computing · Software engineering · Life cycle model · Process model

1 Introduction

The need to access data from any location transformed World Wide Web into an intelligent web which can display quarried data from any location. As the time pass the technologies have improved and need increased day by day. So the experts feel that there is need of some more improved version of business oriented model where hardware, software, tools and applications provided to customer on lease, which can be accessible across the globe via internet. Now this idea of experts and companies implemented through cloud computing where users can get hardware software, tools and applications on lease. Cloud computing has brought the innovative period of potential computing, change a huge branch of IT trade, redesign the buy and use, and get significant consideration worldwide and local society of IT, national and intercontinental organizations [3, 5].

Cloud computing benefits software engineering concepts like agility, availability and cost efficiency. These need to be well engineered for cloud platforms using the software engineering methodologies particularly modeling the cloud aspects to provide logical tested solution prior to implementation to improve the quality. The models assist in representing the problem and its solution (logically) in a methodical mode. They also demonstrate the precise details for different viewpoints and at various stages of development [2]. The goals of modeling and design of cloud paradigm, are to

© ICST Institute for Computer Sciences, Social Informatics and Telecommunications Engineering 2017
J. Ferreira and M. Alam (Eds.): Future 5V 2016, LNICST 185, pp. 189–195, 2017.
DOI: 10.1007/978-3-319-51207-5_19

support all the activities of cloud modeling and design to fit into an overall model-based development from both business and IT perspectives. Unfortunately, there is not yet a consensus on the right set of process models.

The rest of the paper is organized as follow: Sect. 2 illustrates the detail background with regard to the development of the cloud, Sect. 3 introduces the proposed cloud computing development life cycle model, and finally Sect. 4 includes the conclusion.

2 Motivation and Background

Software models have been directing the practice in the domain of software engineering. The process model is actually a set of logically associated activities whose implementation guides to the construction of a specific software product [6]. There are frequently precise specifications and in order to obtain them there may be need to combine different cloud services [6].

Generally, we can have two possible ways to utilize the software engineering in the paradigm of cloud computing; (i) to improve the software development methods to suite designing software applications for the cloud (ii) the utilization of cloud to support the process of software development [1]. Cloud computing benefits software engineering concepts like agility, availability and cost efficiency. These need to be well engineered for cloud platforms using the software engineering methodologies particularly modeling the cloud aspects to provide logical tested solution prior to implementation to improve the quality. Modeling of cloud computing can be used to generate code and offer code generation as a service which is the attractive and demanding activity [4].

There is a huge anxiety that what type of service and requirements of the organization should be moved to the cloud and this migration may take place based on the QoS [8, 9]. To identify and categorize the services of being provided by the cloud, the services should be ranked [11]. The scalability of the cloud should also be kept in mind and should be a very obvious part of the cloud development process because the nodes which are part of the cloud may be located in different regions of the world [10]. The Service Level Agreement (SLA) is also a major activity involved in the cloud computing paradigm, thus a proper management of the SLA is also important [12].

The cloud development is a very sensitive issue in terms of its flexible nature. The development of cloud should be taken into consideration throughout the entire life cycle f the cloud development. Global Software Development could be one of the ideal techniques to utilize for the development of cloud computing [13, 14]. Risk mitigation, monitoring and management activity could play a vital role in the development of cloud.

3 Cloud Computing Development Life Cycle Model (CCDLC)

This section includes the detail description of the proposed life cycle model. Software development model can be different for cloud computing than traditional development because cloud need to consider services to internal consumptions as well as external users. The target market of developing cloud application is small and most of them are not fully aware of the development life cycle on cloud computing. The services to external users the development life cycle can be different because it is integrated with internal consumption of cloud service.

There is need to integrate other significant processes with SDLC, because during cloud development there is need of other processes such as feasibility analysis, risk management, security and privacy checks, scalability, efficiency, SLA and other QoS factors. Therefore, we propose, a specific life cycle model for the cloud development which tackle the above mentioned characteristics of the cloud computing. The abstract description of the proposed life cycle model is depicted in Fig. 1. The figure highlights and visualizes the overall proposed model and depicts all the phases which are proposed as part of the proposed model. The detailed description of each and every phases of the proposed cloud computing life cycle model (CCDLC) is given below:

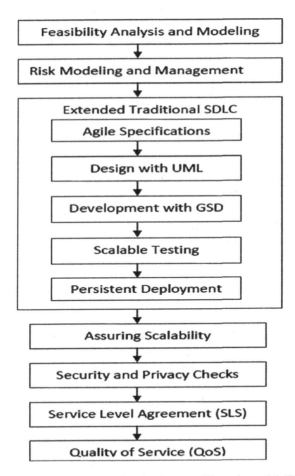

Fig. 1. Proposed cloud computing development life cycle model (CCDLC)

3.1   Feasibility Analysis and Modeling

This phase is about the decision making by the organization to migrate cloud paradigm. Now a days the cloud computing has become the most demanding paradigm and every organization is moving to the cloud. There are a number of reasons for the reputation such as extreme scalability, accessibility and the reduction in the IT cost. However, several organizations still get it tough to migrate their application to the cloud. The complexity of migrating to cloud could be because of different reasons. Therefore, the first step we propose in the proposed model is to clearly take a clear decision to migrate to the cloud. The decision should be based on obvious advantages and pros.

3.2   Risk Modeling and Management

Once the decision is made to migrate to the cloud by organization, and then the risk modeling and its mitigation and management should be taken into account. Risk is the opportunity that an incident will happen and unfavorably influence the results. This phase plays an important role in the overall designing of the cloud. Cloud may affect by a number of risks such as security risks, privacy issue, technical risks, social risks, and accessibility issues. Therefore, these issues are needed to be identified and mitigated properly before going to develop cloud.

3.3   Extended Traditional Software Development Life Cycle (SDLC)

The traditional SDLC include requirement, design, development, validation, and deployment. In the cloud development, these phases are taken into account with some modifications and extensions. The detail description of these phases is given below:

3.3.1   Agile Requirement Engineering

The traditional requirement engineering process can be extended in the way to continuously manage the requirements with the agility. The requirement phase includes the identification and elicitation of the cloud services; it may include the subscription to an existing service in the cloud, developing in-house service using the cloud environment, or outsourcing service.

3.3.2   Modeling and Design Using UML Extension Mechanism

Architecture and design is the next phase of traditional generic SDLC. In this phase the proper architecture (the logical solution) of the system is built and mapped to the requirement. Unified Modeling Language (UML) can be extended to model the cloud aspects in the overall process of the cloud development.

3.3.3   Development with GSD Techniques

The next phase of traditional SDLC is the development. The cloud development is a very challenging and complex task. The traditional SDLC development phase may extend to Global Software Development (GSD) to achieve the required services according the specifications. Now days, we are practicing an explosion of cloud computing as a novel and innovative generation of the internet. Every organization is migrating to the cloud. Therefore, the development of the applications should be taken into consideration with the experience and skills of international experts using the GSD techniques.

3.3.4 Scalable Testing

The next phase of traditional SDLC is the testing. In this phase the developed systems is tested, verified and validated thoroughly. As the cloud is a complex distributed system, therefore, the traditional tasting techniques will be extended to have scalable testing methods and tools to validate and test the cloud services. The scalable testing may include automated testing procedure and automated tools to test the services automatically.

3.3.5 Deployment with Persistency

Deployment is the very last phase of traditional SDLC. Traditional deployment is one big bang activity which is taken place at the end of the development. Cloud is a complex distributed system, therefore, it requires a proper continues deployment process in the form of persistent deployment. Persistent deployment is a kind of extension to continue integration whose purpose is to minimize the load time between the development and deployment. To get the persistent deployment the development squad relies on infrastructure that mechanize and appliance a variety of steps leading up to deployment.

3.4 Assuring Scalability

The capability to scale on demand is the greatest benefit of cloud computing. When taking into consideration the variety of advantages of cloud, scaling on-demand is one of the all time great advantage in the cloud computing paradigm. Thus, after the development of the cloud, the cloud should assure the scalability and it should be scalable to provide the services on demand at any stage. This scaling facility should be auto by the vendor.

3.5 Security and Privacy Checks

Security and privacy have consistently been a big issue in IT applications. In the cloud computing paradigm, it becomes principally severe because the services situated in diverse locations across the globe. Security and privacy are the two major features of user's anxiety in the cloud computing. Therefore, the security and privacy should be handled properly after the development and deployment of the cloud.

3.6 Service Level Agreement (SLA)

Security and privacy have consistently been a big issue in IT applications. In the cloud computing paradigm, it becomes principally severe because the services situated in diverse locations across the globe. Security and privacy are the two major features of user's anxiety in the cloud computing. Therefore, the security and privacy should be handled properly after the development and deployment of the cloud.

3.7   Quality of Service (QoS) Management

Due to the numerous use of cloud computing, the QoS of cloud computing has turn out to be a significant and necessary matter as there are numerous open issues which required to be resolved associated to the trust in cloud computing paradigm. QoS management involves guaranteeing the level of service along with all its attributes such as availability, performance and reliability. In this phase proper QoS modeling approaches are needed to be used assure and manage the QoS.

4   Conclusion

Cloud computing development life cycle (CCDLC) model is proposed for the designing and developing of cloud from both vendor and consumer perspectives. The proposed model overcomes the shortcomings and deficiencies found in the traditional existing software engineering process models. The proposed process model includes the modifications and extension of the traditional SDLC and other important processes.

References

1. Whittle, J., et al.: The state of practice in model-driven engineering. IEEE Softw. **31**(3), 79–85 (2014)
2. Brunelière, H., et al.: Combining model-driven engineering and cloud computing. In: 4th Workshop on Modeling, Design, and Analysis for the Service Cloud, Paris, France, June 2010
3. Armbrust, M., Fox, A., Griffith, R., Joseph, A.D., Katz, R., Konwinski, A., Lee, G., Patterson, D., Rabkin, A., Stoica, I., Zaharia, M.: A view of cloud computing. Commun. ACM **53**(4), 50–58 (2010)
4. Crocombe, R., Kolovos, D.: Code generation as a service. In: Proceedings of ACM/IEEE 18th International Conference on Model Driven Engineering Languages and Systems, Ottawa, Canada, 29 September 2015
5. Silva, G.C., Rose, L.M., Calinescu, R.: Cloud DSL: a language for supporting cloud portability by describing cloud entities. In: Proceedings of the 2nd International Workshop on Model-Driven Engineering on and for the Cloud co-located with the 17th International Conference on Model Driven Engineering Languages and Systems (MoDELS 2014), Valencia, Spain, 30 September 2014
6. Sommerville, I.: Software Engineering, 9th edn. Addison-Wesley, Boston (2011)
7. Patidar, S., Rane, D., Jain, P.: Challenges of software development on cloud platform. In: 2011 World Congress on Information and Communication Technologies (WICT), pp. 1009–1013 (2011)

8. Saripalli, P., Pingali, G.: MADMAC: multiple attribute decision methodology for adoption of clouds. In: 2011 IEEE International Conference on Cloud Computing (CLOUD), 4–9 July 2011, pp. 316–323 (2011)
9. Beserra, P.V., Camara, A., Ximenes, R., Albuquerque, A.B., Mendonca, N.C.: Cloudstep: a step-by-step decision process to support legacy application migration to the cloud. In: 2012 IEEE 6th International Workshop on Maintenance and Evolution of Service-Oriented and Cloud-Based Systems (MESOCA), 24–24 September 2012, pp. 7–16 (2012)
10. Gibson, J., Rondeau, R., Eveleigh, D., Tan, Q.: Benefits and challenges of three cloud computing service models. In: 2012 Fourth International Conference on Computational Aspects of Social Networks (CASoN), 21–23 November 2012, pp. 198–205 (2012)
11. Garg, S.K., Versteeg, S., Buyya, R.: A framework for ranking of cloud computing services. Future Gener. Comput. Syst. **29**(4), 1012–1023 (2013). http://dx.doi.org/10.1016/j.future.2012.06.006
12. Undheim, A., Chilwan, A., Heegaard, P.: Differentiated availability in cloud computing SLAs. In: 2011 12th IEEE/ACM International Conference on Grid Computing (GRID), pp. 129–136 , September 2011
13. Smirnova, I.: Impact of cloud computing on global software development challenges. In: Proceedings of Cloud-Based Software Engineering, pp. 71–78. University of Helsinki, Helsinki (2013)
14. Haig-Smith, T., Tanner, M.: Cloud computing as an enabler of agile global software development. Issues Inf. Sci. Inf. Technol. **13**, 121–144 (2016). http://www.informingscience.org/Publications/3476

Effective Use of Search Functionality in Pakistan E-Government Websites

Mutahira Naseem Tahir[(⊠)] and Fahim Arif

Military College of Signals,
National University of Sciences and Technology (NUST), Islamabad, Pakistan
mutahiranaseem.mcscsl9@students.mcs.edu.pk,
fahim@mcs.edu.pk

Abstract. The modern era of technology makes technology an integrated part of our lives. The information and communication technology ICTs and information technology has diverse impacts on the human lives both in positive and negative aspects. The modern emerging trends in the technology and their impact on the human race encourage researchers to develop new enhanced techniques. The researchers proposed the concept of e-government to use the information technology and information and communication technology (ICT) and web-services to encourage the involvement of different stakeholders in the development and enhancement of government procedures. The electronic government directorate of Pakistan under the supervision of Ministry of Information approves an e-government plan. This paper provides an insight into the importance of search functionality in the e-government websites of Pakistan. A survey is conducted to evaluate the importance of search functionality and effective use of search functionality to improve the effectiveness, efficiency and satisfaction of the websites.

Keywords: E-government · Search functionality in e-government · Effective use of search functionality

1 Introduction

The basic definition of an E-Government is "The employment of the Internet and the world-wide-web for delivering government information and services to the citizens" [1]. Electronic government can also be stated as "The utilization of Information Technology (IT), Information and Communication Technologies (ICTs), and other web-based telecommunication technologies to improve and/or enhance on the efficiency and effectiveness of service delivery in the public sector" [2]. E-government encourages the involvement of different stakeholders in the development of state and community as well as develops the governance process better [3].

Network society theorists' states in various studies that activities and social structures in societies are increasingly centers around network largely on information and communication technologies (ICTs) [4–6]. Though most governments hesitates in taking up new technologies. The current practices of public administrations become increasingly complex, have to harmonize activities the developing stakeholders in the

© ICST Institute for Computer Sciences, Social Informatics and Telecommunications Engineering 2017
J. Ferreira and M. Alam (Eds.): Future 5V 2016, LNICST 185, pp. 196–204, 2017.
DOI: 10.1007/978-3-319-51207-5_20

public domain. The management and processing of huge amount of information seems like a silo, insular culture, slow decision making and information dispersion of the old model appears to be not suited for the advance information dissemination and corporation, legitimacy and trust professed by the citizens and eventually affects efficiency and proficiency [7].

Since the middle of 1990s, the e-government and ICT become the ideal solutions for the traditional public administration problems. Researchers are of opinion that intensive use of technology could increase the efficiency of public administration operational rules, abridges organizational procedures [8], citizen's involvement to be increased [9] and transparent and accountable government activities could be legislate [10]. High expectations requires an exploration to evaluate the limit to which public administration and procedures can integrate ICTs in their basic activities. It is fascinating to validate whether governments are becoming suppler, transparent and considers citizen's preferences, and transition towards new form of administrative organization might be considered as a virtual state [11] or as a network administration [12].

The government of Pakistan announced its first e-government strategy in 2005 for five years. The E-Government Directorate revised the strategic planning based on the consideration of past years, new realities and current conditions. The E-Government strategic plan is approved for three years in July 2012 [13]. Salient feature of this strategic plan is human resource development, top-level owner-ship, comprehensive planning, priority on high impact projects, interoperability of applications, security of government information, timely availability of funds and software development. The Government of Pakistan aims to spend Rs.4.6 billion on the information technology in the form of e-government, infrastructure and human resource development. The e-government is gaining fame in recent years. It is observed that over the past few years, almost every ministry is moving towards the e-government. This allows the citizens of Pakistan to have an easy access to rules, regulations, legislations and laws. The e-government sector is in its initial stages having many short comings. Apart from many other limitations, these government sector websites do not address the issue the intelligent information retrieval.

This paper is organized in section; in Sect. 2 literature review is provided, Sect. 3 provide the evaluation of effective use search functionality and Sect. 4 provides with the conclusion.

2 Literature Review

The traditional web applications do not address the features and usage context of mobile applications. In the paper [18], authors highlight the issues of user interface operability and their impact on the evaluation on the mobile applications usability. An ISO 25010 based quality model called 2Q2U is offered to assess and improve the MobileApp usability. 2Q2U addresses the quality attribute (1) quality, (2) quality in use, (3) actual usability and (4) user experience. They looked as the mobile app usage context and impact on app design. In combination with 2Q2U a new framework for the quality is proposed in [19]. SIQinU (strategy to improve quality in use) aims to improve the quality. The propose strategy is used for the evaluation of radio WebApps for user's hedonic and pragmatic needs.

A framework is presented [20] to evaluate Hindi and Punjabi websites on the basis of external quality attributes based on ISO/IEC 25010-11 standard. The proposed framework includes visual and automated observations. In their research proposal, 290 websites from academic, newspaper and government domains will be selected to evaluate on the basis of designed framework.

Abbasi et al. in [21] presents the ISO 25010 user interface aesthetics (UIA) from product quality (PQ) and quality in use (QIU) perspectives. In the said ISO standard, aesthetics is categorized as sub-characteristics of usability. While designing, it is important to design and evaluate aesthetics from user's perspective as well as from the product design perspective. The proposed approached is used to evaluate local Chinese websites.

The web portals as discussed in [22] provide wide range of applications, information and services. The quality in use, as important perspective of ISO/IEC 25010 standard is used as base to define a quality model for the web portals.

Stefani in [23] defines the B2C-specific quality assessment model for web metrics. Three dimensions based on end-user interaction categories, internal specifications of metrics and sub-characteristics of quality in light of ISO9126.

Jayakumar [24] evaluates the websites quality on the basis of accuracy, feasibility, usability and Website quality Assessment model (WQAM). The quality metrics are evaluated through the questionnaire sample. The feedback is used to identify the areas in the website requires improvement. On the basis of the feedback, new e-learning framework is proposed to incorporate the findings. High-level structure based on characteristics, sub-characteristics and main three dimensions attributes (content, service and technical quality) is suggested for the comparison, evaluation and improvement of websites [25].

The independency between the six quality attributes of ISO/IEC9126 is confirmed in the paper [26]. Customer satisfaction is quantitatively measured on the basis of these quality attributes. A prediction model is presented to assess the total customer satisfaction based on the inherit characteristics of the product. Based on these quality attributes "three dimensional integrated value model [27] " model is proposed to display the total quality of the system visually. The entire quality of the system can be measured and compared using integration meter of the evaluation of the system presented through solid volume cubic vectors of the characteristics.

3   Results

The intended context of use of the e-government websites is describe in this section. This section describes the participants who took part in the quality in use evaluation process.

3.1   Effectiveness in Use

The data for the Effectiveness in use was collected using the usability test, observations, retrospective think aloud (RTA) and post evaluation questionnaire. The results for effectiveness were recorded for all the eight websites during the task performance phase. The mean, median and mode values for each websites is shown in Table 1.

Table 1. Results for effective in use

Sr. No.	Websites	Mean %age completion rate			User response		
		Mean	Median	Mode	Mean	Median	Mode
1.	FBR	70%	67%	60%	2.355	2	2
2.	Ministry of Defence	56.66%	49%	50%	2.5	1	2
3.	Federal education Institute Directorate	46.66%	35%	40%	1.3	1	1
4.	Ministry of Finance	63.33%	55%	50%	2	2	2
5.	Ministry of Professional and Technical Training	53.33%	60%	45%	2.3	1	2
6.	Ministry of Planning, Development and Reform	46.67%	38%	48%	3	1	2
7.	Official Gateway to Government of Pakistan	40%	25%	35%	1.3	2	1
8.	Ministry of Science and Technology	54%	56%	45%	2	2	2

Note: all task complete = 100%, each task with error = 10% deduction, help from evaluator = 5% deduction User Response: 1 = strongly not agree, 2 = disagree, 3 = neutral/somewhat agree, 4 = agree, 5 = strongly agree

The analysis of the Fig. 1 showed that measures for the website of FBR are greater than all other websites. This shows that users find it easy to complete the assigned tasks with less errors and least help from the evaluator. The results indicated that users had more difficulty in completing the tasks, made mistakes and required frequent help from the evaluator while using other websites. The Fig. 1 show that users faced great difficulty in finding the desired results while using the Official Gateway to Government of Pakistan.

The lower scores show poor performance of the users. The user response for Official Gateway to Government of Pakistan and Federal Education Institute

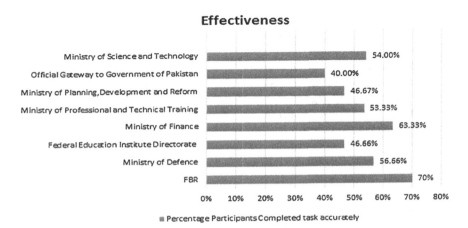

Fig. 1. Comparative analysis of effectiveness in use for the websites

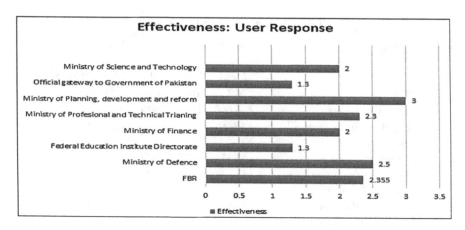

Fig. 2. Effective in use: user response

Directorate is least. The analysis of the results shows users were not or somewhat satisfied with their task performance as shown in Fig. 2.

3.2 Efficiency in Use

The mean time taken to complete the accurately completed task is measures using the usability test and observations. The mean, median and mode values measures through the post evaluation satisfaction questionnaire for each websites is shown in Table 2.

Table 2. Efficiency in use

Sr. No	Websites	Mean efficiency			User response		
		Mean %age task rate (%age)	Time	%age / Time	Mean	Median	Mode
1.	FBR	70%	1.25 m	56%	2.473	1	2
2.	Ministry of Defence	56.66%	1.5 m	37.77%	2.7	2	2
3.	Federal education Institute Directorate	46.66%	2.5 m	19%	1	1	1
4.	Ministry of Finance	63.33%	2.5 m	25.33%	2.1	1	2
5.	Ministry of Professional and Technical Training	53.33%	1.25 m	42.66%	1	2	1
6.	Ministry of Planning, Development and reform	46.67%	1.5 m	31.11%	2	2	2
7.	Official Gateway to Government of Pakistan	40%	2.5 m	16%	2.3	1	2
8.	Ministry of Science and Technology	54%	1.12 m	48.21%	1	1	1

Note: Mean %age task rate (%age) is used from the Table 4.1, Time is calculated in minutes using stop watch, User Response: 1 = strongly not agree, 2 = disagree, 3 = neutral/somewhat agree, 4 = agree, 5 = strongly agree

The comparative analysis of the results shown as bar graph in Fig. 3. The analysis of the results shows that the efficiency rate of the selected websites were not high. These websites do not provide functions that access users in finding the desired results within specified time. Other factor that affect the efficiency was the organization of the data on these websites. Users found it difficult to memorize these websites for later use. The website of FBR presented the partial search functionality helped in achieving high efficiency rate. On the other hand Official Gateway to Government of Pakistan and Federal Education Institute Directorate had the least efficiency rates.

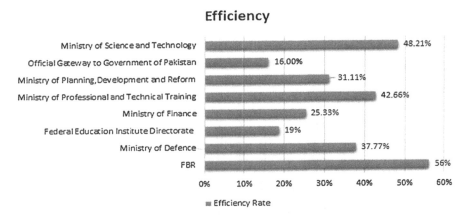

Fig. 3. Efficiency

The high scale indicate the greater satisfaction achieved obtained by completing tasks in limited/ minimum time. The users did not feel satisfied after achieving the goals shown in the Fig. 4.

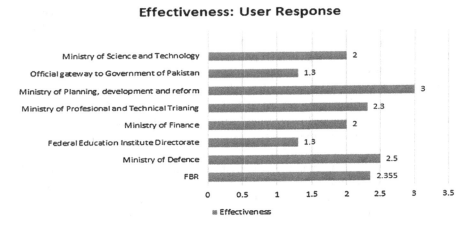

Fig. 4. Efficiency: user response

3.3   Satisfaction in Use

Satisfaction is accomplished by the achieving the pragmatic goals. Satisfaction in use in measured in terms of satisfaction, likability, trust, pleasure and comfort obtained while using the system. The response showed users were not felt accomplished after achieving the tasks. The pragmatic goal did not fulfilled and users faced number of difficulties in achieving the goals (Fig. 5).

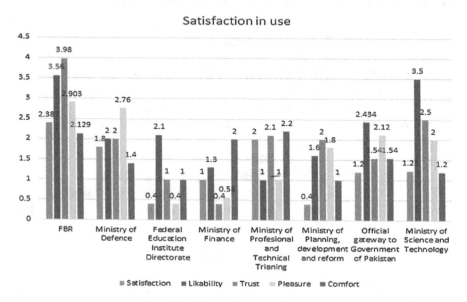

Fig. 5. Satisfaction in use

4   Conclusion

For the survey eight different websites were selected. The criteria for the selection was (1) information on the website, (2) number of pages, (3) website layout, (4) importance of data and (5) amount of data on the website. These factors has great impact on the learnability, cognitive learning, though process, likability, pleasure, comfort and trust of the users on these websites. An important factor that contribute to poor performance is lack of search feature in those websites. The tasks for the survey were organized in such a way that participants were asked to use the websites for 20–30 min to get familiar with the features of the websites. After the said time period, different tasks were assigned to them. Most tasks were designed to find a particular data present on these websites. The users found it hard to memorize the provided information leads to poor performance in completing the task accurately within the allocated time period. The satisfaction in use directly associated with the effectiveness in use and efficiency in use. If the user fails to complete the tasks accurately without external help within the specified time leads to poor satisfaction results. The trust, comfort, pleasure, likability and satisfaction are associated with the pragmatic goals of the users. The users faced difficulty in using these websites, find it hard to achieve the desired goals. The results shows that users had to explore the websites multiple times before reaching to the destination. Due to lack of search feature, user's performance degraded. The participants did not perform their tasks with effectiveness and efficiently. Once the tasks not completed effectively and efficiently, user satisfaction, trust, comfort and pleasure had not achieved. The users faced difficulty in using these websites, find it hard to achieve the desired goals. The results shows that users had to explore the websites multiple times before reaching to the destination.

References

1. United Nations E-Government Survey 2014. United Nations Department of Economic and Social Affairs, UN. Accessed 16 Sept 2014
2. Hai, J.C.: Ibrahim. Fundamental of Development Administration. Scholar Press, Selangor (2007). ISBN 978-967-5-04508-0
3. Alenezi, H., Tarhini, A., Sharma, S.K.: Development of quantitative model to investigate the strategic relationship between information quality and e-government benefits. Transforming Gov.: People Process Policy 9(3), 324–351 (2016). doi:10.1108/TG01-2015-0004. Accessed 5 Jan 2016
4. Castells, M.: La Era de la Información (trilogy) (The Information Age). In: Economy, Society and Culture Mexico XXI Century 1996 Alianza, vol. 1, Madrid (1996–2000)
5. Dijk, J.V.: The Network Society Social Aspects of New Media. SAGE Publications Ltd., Thousand Oaks (1999)
6. Wall-Smith, M.: The network society: a shift in cognitive ecologies? First Monday 7 (2002)
7. Schellong, A.: Crossing the boundary - why putting the e in government is the easy part. In: PNG Working paper N. PNG07–002, John F. Kennedy-Harvard School of Government, (2007)
8. Fountain, J.: Challenges to organizational change: multi-level integrated information structures (MIIS). In: Lazer, D., Mayer-Schoenberger, V. (eds.) Governance and Information Technology: From Electronic Government to Information Government. MIT Press, Cambridge (2007)
9. Hague, B.N., Loader, B.D.: Digital Democracy: Discourse and Decision Making in the Information Age. Routledge, London (1999)
10. Bhatnagar, S.: Access to information: e-government. In: Hodess, R., Inowlocki, T., Wolfe, T. (eds.) Global Corruption Report. Profile Books Ltd., London (2003)
11. Fountain, J.: Building the Virtual State: Information Technology and Institutional Change. Brookings Institution Press, Washington, D.C. (2001)
12. Castells, M., Ollé, E.: El model Barcelona II: L'Ajuntament de Barcelona a la societat xarxa, model II Barcelona: Barcelona city council in the network society in Catalonia (2002–2004). Universitat Oberta de Catalunya, Barcelona (2004)
13. E-government strategic plan for the federal government, Electronic Government Directorate - Ministry of Information Technology, July 2012
14. Internet live stats, 10 May 2016. http://www.internetlivestats.com/total-number-ofwebsites/
15. Internet service providers association of Pakistan, 20 May 2016. http://www.ispak.pk/index.php
16. State of social media in Pakistan in 2016, 10 May 2016. http://propakistani.pk/2016/01/26/state-of-social-media-in-pakistan-in2016/

17. Education, 10 May 2016. http://finance.gov.pk/survey/chapter_10/10_Education.pdf
18. Lew, P., Olsina, L.: Towards understanding and improving mobile user experience. In: Proceedings of PNSQC (2013)
19. Lew, P., Abbasi, M.Q., Rafique, I., Wang, X., Olsina, L.: Using web quality models and questionnaires for web applications understanding and evaluation. In: Proceedings of the 2012 Eighth International Conference on the Quality of Information and Communications Technology, 02–06 September, pp. 20–29 (2012)
20. Kaur, R., Goyal, V., Kaur, G.: Web quality model for websites developed in Punjabi and Hindi. Int. J. Soft Comput. Softw. Eng. [JSCSE] 3(3), 557–563 (2013). Special Issue: The Proceeding of International Conference on Soft Computing and Software Engineering 2013 [SCSE 2013], San Francisco State University, CA, USA, March 2013
21. Abbasi, M.Q., Weng, J., Wang, Y., Rafique, I., Wang, X., Lew, P.: Modeling and evaluating user interface aesthetics employing ISO 25010 quality standard. In: 2012 Eighth International Conference on Quality of Information and Communications Technology (QUATIC), Lisbon, pp. 303–306, 3–6 September 2012
22. Herrera, M., Moraga, M.Á., Caballero, I., Calero, C.: Quality in use model for web portals (QiUWeP). In: Daniel, F., Facca, F.M. (eds.) ICWE 2010. LNCS, vol. 6385, pp. 91–101. Springer, Heidelberg (2010). doi:10.1007/978-3-642-16985-4_9
23. Stefani, A., Vassiliadis, B.: A web metrics quality evaluation framework for e-commerce systems. In: Proceedings of the ICWE 2005 Conference, pp. 110–123 (2005)
24. Jayakumar, R., Mukhopadhyay, B.: Website quality assessment model (WQAM) for developing efficient e-learning framework-a novel approach. Int. J. Eng. Technol. (IJET) 5, 3370–3380 (2013)
25. Rocha, A.: Framework for a global quality evaluation of a website. Online information review. Int. J. Digit. Inf. Res. Use 36, 374–382 (2012)
26. Esaki, K.: Prediction models for total customer satisfaction based on the ISO/IEC9126 system quality model. Am. J. Oper. Res. 3, 393–401 (2013)
27. Esaki, K.: Three dimensional integrated value models based on ISO/IEC9126 system quality model. Am. J. Oper. Res. 3, 342–349 (2013)

Ready Mealy, Moore & Markov Mathematical Modeling Machines for Big Data

Hafiz Tahir Furqan, Ahmad Abdul Wahab,
Sultan Sadiq, and Pir Amad Ali Shah[✉]

Computer Science Department, University of South Asia, Lahore, Pakistan
hmftj@live.com, abdulwahabahmad96@gmail.com,
shksadiqsultan@gmail.com, amad.ali@usa.edu.pk

Abstract. In this paper we have given a new idea of bringing automation to the Big data. This paper basically focused on the three core topics in automaton for Big data. Automation machines are used in today world to automate certain applied sciences ideas, into computational models. Mealy, Moore & Markov Probabilistic Modeling are one of those used to accomplish the said task. In particular we have focused on transducer and features in MATLAB and implication of system General Knowledge are focused in this paper.

Keywords: Markov model · Probability · Hidden-Markov model · Finite state machine · Process control · Finite state automaton · Big data

1 Introduction

A mathematical framework for the modeling, trying out and diagnosis of sequential machines is developed. A completely standard system model is used in which a transition gadget is represented as a sequential system, probably with state and output unit one-of-a-kind from the ones of the best system. A deterministic finite automaton, called observer, describes the technique by way of which one profits information from the statement of the responses to test sequences. It generalizes the paintings of Henie on distinguishing and homing sequences, through modeling all of the possible conclusions that might be drawn from gazing the circuit beneath take a look at. A nondeterministic acceptor is derived from the observer; it accepts diagnosing sequences and can also be used to generate test sequences. We then accomplice possibilities with this nondeterministic acceptor which, collectively with a stochastic supply of enter symbols, provides a probabilistic prognosis. As a selected utility we do not forget the checking out and prognosis of random-get right of entry to reminiscences via random take a look at sequences. The model in [1] generalizes the paintings by means on the calculation of the period of a random check collection required to assure that the chance of detection of a fault exceeds a prescribed threshold [2].

© ICST Institute for Computer Sciences, Social Informatics and Telecommunications Engineering 2017
J. Ferreira and M. Alam (Eds.): Future 5V 2016, LNICST 185, pp. 205–214, 2017.
DOI: 10.1007/978-3-319-51207-5_21

2   Moore Machine Background

The interesting records of the way finite automata have become a branch of laptop technological know-how illustrates its wide range of programs. the first humans to don't forget the concept of a finite-nation device protected a team of biologists, psychologists, mathematicians, engineers and some of the first laptop scientists. They all shared a common interest: to model the human concept procedure, whether within the brain or in a laptop. Warren McCulloch and Walter Pitts, neurophysiologists, were the first to present an outline of finite automata in 1943. Their paper, entitled, "A Logical Calculus Immanent in apprehensive hobby", made great contributions to the look at of neural network concept, concept of automata, the concept of computation and cybernetics. Later, two laptop scientists, G.H. Mealy and E.F. Moore, generalized the concept to a great deal greater effective machines in separate papers, posted in 1955-fifty six. The finite-state- machines, the Mealy system and the Moore gadget, are named in recognition of their paintings, while the Mealy system determines its outputs thru the present day state and the center, the Moore device's output is based upon the modern nation alone.

2.1   Moore Machine Introduction

Moore machine is an FSM whose outputs depend on only the present state. A limited state machine (FSM) or limited robotic (FSA, plural: automata), or basically a country system, is a numerical version of calculation used to define both laptop applications and consecutive motive circuits. It is taken into consideration as a unique device that may be in considered one of a constrained variety of states. The device is in one and simplest nation without delay; the state- its miles in at any given time is called the existing nation. It could change starting with one state- then onto the following whilst began with the aid of an activating occasion or situation; that is called a flow. A selected FSM is characterized by using a rundown of its states, and the activating situation for every circulation. The behavior of country machines may be visible in numerous devices in reducing edge society that perform a foreordained arrangement of activities depending upon a succession of occasions with which they're displayed. trustworthy instances are sweet machines, which apportion items when the great feasible blend of cash is kept, lifts, which drop riders off at top floors earlier than happening, pastime lighting, which alternate grouping whilst automobiles are maintaining up, and blend locks, which require the contribution of mix numbers in the perfect request. Restrained nation machines can reveal a big number of problems, amongst which might be digital configuration computerization, correspondence convention plan, dialect parsing and different building programs. The FSM memory is restrained through the quantity of states.

Moore gadget is an FSM whose outputs rely upon handiest the existing nation [6]. A Moore device may be defined by means of a 6 tupple (Q, Σ, O, δ, X, q0) where

- Q is a finite set of states.
- Σ is a finite set of symbols referred to as the enter alphabet.

- Is a finite set of symbols known as the output alphabet?
- δ is the input transition function where δ: $Q \times \Sigma \rightarrow Q$
- X is the output transition characteristic wherein X: $Q \times \Sigma \rightarrow O$
- Q0 is the preliminary country from where any enter is processed ($q0 \in Q$).

The crucial distinction is that there may be no arrangement of definite states, and that the circulate capacity places you in every other state, in addition to produces a yield picture. The goal of this form of FSM isn't tolerating or dismissing strings, but instead producing an association of yields given an association of inputs that a discovery takes in inputs, bureaucracy, and creates yields. FSMs are one technique for portraying how the inputs are being treated, deliberating the inputs and nation, to create yields. On this way, we are extraordinarily inspired by using what yield is created. In DFAs, we could not care less what yield is created. We thought simply whether a string has been recounted via the DFA or no longer.

2.2 Moore Design

The basic operation of a state-machine has the following two properties in [7]

- It traverses thru a series of the states, where the subsequent state- is determined by subsequent state decoder, depending upon the present state and input situations.
- It gives sequences of output alerts primarily based upon state- transitions. The outputs are generated by the output decoder based upon present country and enter conditions.

The use of input signals for finding out the next country is likewise known as branching. Similarly to branching, complicated sequencers provide the functionality of repeating sequences (looping) and subroutines. The transitions from one country to another are called manipulate sequencing and the common sense required for determining the following states is called the transition function. The usage of enter alerts within the decision-making manner for output generation determines the type of a nation machine. There are well known sorts of nation machines: Mealy and Moore. Moore state system outputs are a feature of the prevailing country simplest. Such machines are fantastically susceptible to hazards, difficult to design and are seldom used. In our discussion we will attention completely on sequential state machines (Figs. 1 and 2).

The capabilities carried out the entire system layout functions performed via controllers can be categorized as one of the following nation system capabilities:

- Arbitration
- Occasion monitoring
- A couple of condition checking out
- Timing delays
- Manage sign era

Later we will take a layout instance and illustrate how these functions can be used whilst designing a nation system. State-system principle supply us a hazard to analyze

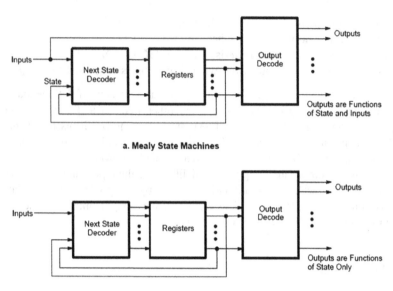

Fig. 1. The two standard state machine model

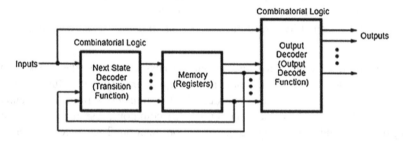

Fig. 2. State machine with separate output and next state decoder

the fundamental speculation for all successive reason frameworks, the confined country system (FSM), on the other hand essentially state machine. Those parts of automatic frameworks whose yields depend on upon their beyond inputs and moreover their gift ones can be displayed as constrained state- machines. The "records" of the device is summed up in the estimation of its inner nation. At the point when every other information is exhibited to the FSM, a yield is produced which relies on upon this statistics and the present situation of the FSM, and the device is brought on to transport into new country, alluded to as the subsequent nation. This new state- likewise relies on upon both the statistics and gift country. The structure of a FSM is verified pictorially. The interior country is positioned away in a chunk marked "reminiscence." As talked about before, two combinatorial capacities are required: the move capacity, which creates the estimation of the following state-, and the yield capacity, which produces the country machine yield.

2.3 Moore Review

A state is a portrayal of the status of a framework that is holding up to execute a move. A move is an arrangement of activities to be executed when a condition is satisfied or when an occasion is gotten. For instance, when utilizing a sound framework to listen to the radio (the framework is in the "radio" state), getting a "next" boost results in moving to the following station. At the point when the framework is in the "Compact disc" express, the "following" boost results in moving to the following track. Indistinguishable jolts trigger diverse activities relying upon the present state.

Limited state machine are one method for portraying the conduct of a circuit with state. Consider it an exceptionally unrefined programming dialect, which takes inputs, and uses those inputs and the state to register yields, furthermore to figure out what state to move into. CPUs use limited state machines as control units to synchronize the get, execute, and disentangle cycle. These machines can be fairly modem, be that as it may, programs exists to change over the limited state machine into genuine flip failures and rationale entryways [3]

3 Mealy Machine Introduction

A Mealy machine (S, f) consists of a set S of states and a transition function f: S $\rightarrow$ (B × S) A assigning to each states $\in$S and input symbol a $\in$ A a pair (b, s), consisting of an output symbol b $\in$ B and a next states' $\in$S. Typically one writes. This study plans to fill this gap by enabling the inference of nondeterministic models for black-box reactive systems. The core of our contribution is the algorithm N∗, an extension and systematization of works by [8] to infer nondeterministic Mealy machines. We have conducted an experimental campaign to evaluate N∗ considering various features of the target machines. In the real case, multiple queries and approximate equivalence checks are required instead, causing a decrease in performances that we can assess in a quantitative way. As a further assessment of practical feasibility, we have evaluated N∗ on a working implementation of a TFTP client/server protocol [9].

3.1 Mealy State Machine Review

The Mealy state machine design is described in the following stages

- Identify state variables S.
- Identify output decoder and Next state decoder.
- Build state transition diagram.
- Minimize states.
- Choose appropriate type of flip flops.
- Choose state assignment (Assignment of Binary codes to machine states).
- Design next state decoder and output decoder-Use combinational logic structured design methods.

We have given a co-algebraic account of Mealy machines and provided a logical specification language for them. Despite its simplicity, the logic is expressive in the sense that all Mealy machines can be characterized by finite formulae, but also in the sense that logical equivalence corresponds to bi-simulation. Further, the logic is sound and the modal fragment complete for all Mealy machines (Figs. 3, 4 and Table 1).

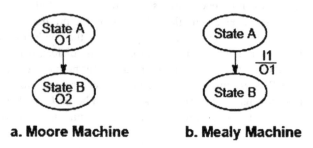

a. Moore Machine   **b. Mealy Machine**

Fig. 3. Output generation in both machines

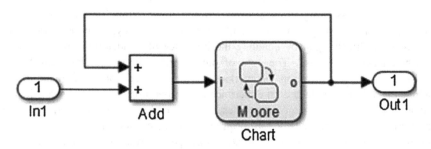

Fig. 4. Moore machine input-output and address operation

Table 1. Mealy Moore machine characteristics

Mealy machine	Moore machine
• Output depends both upon present state and present input	• Output depends only upon the present state
• By and large, it has fewer states than Moore Machine	• For the most part, it has more states than Mealy Machine
• Yield changes at the dock edges	• Info change can bring about change in yield change when rationale is finished
• Mealy machines react faster to inputs	• In Moore machines, more logic is needed to decode the outputs since it has more circuit delays

4 Proposed Methodology

First of all we search multiple topic related to our research on internet and with discussion with fellows and faculty. Then we decide our title as 6 m's in automata (Mealy Moore Markov Model Mathematical Machine). In this research we have three core topics has mealy Moore & Markov probabilistic model. Basically the first two machines are type of transducer. Transducer are basically defined as an automaton that produces output based on current input and previous state is called transducer. It is of two types: Mealy machine the output depends upon only current state. Moore machine the output depends upon both current state and resultant state. Then we discuss NDFA to DFA conversion. DFA minimization mealy Moore machine with 6 tuple then comparison and differences between both. We discuss Algorithms 4 and 5. Translation from both machines. And at the end we discuss Markov probabilistic model and Hidden Markov Model (Figs. 5 and 6).

Theorem 1 (Gudder [5]). Applying TOM $E \in \Gamma M$, N(H1, H2) on a vector state α ΔN(H1) produces vector state $\beta = E(\alpha) \in \Delta M(H2)$ where $\alpha = [\alpha 1, \alpha 2, ..., \alpha n]T$, αi $\Omega \leq (H1)$, where $\beta = [\beta 1, \beta 2, ..., \beta m]T$, $\beta i \in \Omega \leq (H2)$, and $E \in \Gamma M$, N(H1, H2), and in the following way $\beta i = PN j = 1 Eij(\alpha j)$.

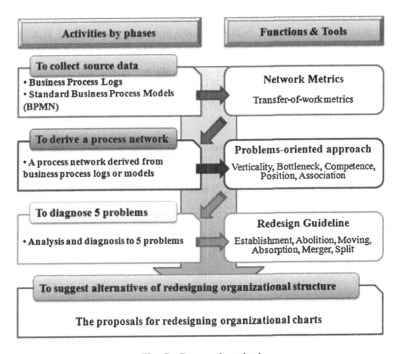

Fig. 5. Proposed method

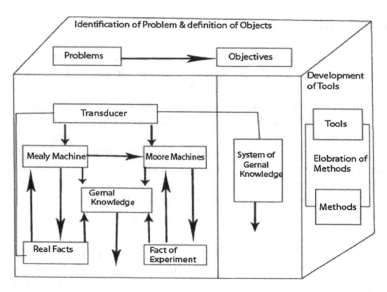

Fig. 6. Transducer derivation during methodology designing.

Theorem 2 (Gudder [5]). Product of TOM A ∈ ΓM, N(H1, H2) and B ∈ ΓN, K (H2, H3) is a TOM ΓM, K(H1, H3) ∋ C = BA.

(Product of two sub-toms is a sub-TOM) Product of sub-toms A ∈ ΓM, N ≤ (H1, H2) and B ∈ ΓN, K ≤ (H2, H3) is a sub-TOM ΓM, K ≤ (H1, H3) ∋ C = BA.

Proof 1 (Lemma 1). According to proof of Lemma 2.2 in [6], Cij = bijaij is a completely positive map. For every ρ ∈ Ω(H1) and j we have that σ = PM i = 1 Aij(ρ) ∈ Ω ≤ (H2). If tr(σ) > 0 then σ̃ = σ/tr(σ) ∈ Ω(H2) and

$$\text{tr}(\sum_{i=1}^{M} \mathcal{B}_{ij}(\sigma)) = \text{tr}(\text{tr}(\sigma) \sum_{i=1}^{M} \mathcal{B}_{ij}(\tilde{\sigma}))$$

$$= \text{tr}(\sigma)\text{tr}(\sum_{i=1}^{M} \mathcal{B}_{ij}(\tilde{\sigma})) \leq 1.$$

In the case where tr(σ) = 0, the σ is the zero operator and PM i = 1 Bij(σ) is also the zero operator. Thus tr PM i = 1 Bij(σ) = 0. Hence, PM i = 1 Ci, j(ρ) ∈ Ω ≤ (H3) and C ∈ ΓM, K ≤ (H1, H3).

Product of (sub-) toms that have same dimensions is associative. (EF)G = E(FG) and (EF)(α) = E(F(α)).

5 Probabilistic Logic

The theory of probabilistic logic has been fully developed in the last two decades. Utley invented a conditional probability computer as early as 1958 (24). The major drawback of his design was that in order to classify an input of n binary items, the number of neurons had to be exponential 2n. It took quite a while to solve this problem and to see the connection of probabilistic logic to probability theory.

Extend the concept of Quantum Markov chains [S. Gudder. J. Math. Phys., 49(7), 2008] for you to advise Quantum Hidden Markov models (QHMMs). For that, we use the notions of Transition Operation Matrices (TOM) and Vector States, which might be an extension of classical stochastic matrices and opportunity distributions. Our fundamental result is the Mealy QHMM components and proofs of algorithms wished for utility of this model: ahead for fashionable case and Vitterbi for a restrained elegance of QHMMs

The problem of the exponential explosion has been solved in the 80s. For singly connected Bayesian networks exact inference is possible in one sweep of Pearl's belief propagation algorithm. A very interesting extension for incomplete data is done by the maximum entropy principle (24.23). This theory can be seen as a realization of von Neumann's prophesied. Probabilistic logic is now used in many fields. To give just one example. I have applied Bayesian networks to population based global optimization (23) [10].

6 Conclusion

Finite state machines are one way of describing the behavior of a circuit with state. Think of it as a very crude programming language, which takes inputs, and uses those inputs and the state to compute outputs, and also to determine what state to transition into. We have got were given added a new model of Quantum Hidden Markov fashions based on the notions of Transition Operation Matrices and Vector States. we additionally proposed a system of the ahead set of guidelines this is applicable for desired QHMMs. CPU's use finite state machines as control units to synchronize the fetch, execute, decode cycle. These machines can be rather sophisticated, however, programs exists to convert the finite state machine into actual flip-flops and logic gates.

References

1. Ampadu, C.: Averaging in SU(2) open quantum random walk. Chin. Phys. B **23**(3), 030302 (2014)
2. Brzozowski, J.A., Jürgensen, H.: A model for sequential machine testing and diagnosis. J. Electron. Test. **3**(3), 219–234 (1992)
3. Knowledge Resouce Library GITAM University. http://www.gitam.edu/library_aboutus. aspx?title=573
4. Ghahramani, Z.: An introduction to hidden Markov models and Bayesian networks. Int. J. Pattern Recogn. Artifi. Intell. **15**(01), 9–42 (2001)
5. Gudder, S.: Quantum Markov chains. J. Math. Phys. **49**(7), 072105 (2008)
6. Automata Theory Tutorial Point Simply Easy Learning. http://www.tutorialspoint.com/ automata_theory/automata_theory_tutorial.pdf
7. Deasing of a Mealy Machine. www.ele.ufes.br

8. Khalili, A., Tacchella, A.: Learning nondeterministic mealy machines. JMLR: Workshop Conf. Proc. **34**, 109–123 (2014)
9. Amato, C., Bonet, B., Zilberstein, S.: Finite-state controllers based on mealy machines for centralized and decentralized POMDPs. In: AAAI, July 2010
10. Fraunhofer Institute for Intelligent Analysis and Information Systems IAIS. https://www.iais.fraunhofer.de/

Education and Socio Economic Factors Impact on Earning for Pakistan - A Bigdata Analysis

Neelam Younas[1,2(✉)], Zahid Asghar[1,2], Muhammad Qayyum[1], and Fazlullah Khan[3]

[1] Pakistan Institute of Development Econonics, Islamabad, Pakistan
qauidian2006@yahoo.com, g.zahid@gmail.com,
qayyum2494@gmail.com
[2] Department of Statistics, Quid-e-Azam University, Islamabad, Pakistan
[3] Department of Computer Science, Abdul Wali Khan University Mardan,
Mardan, Pakistan
fazlullah@awkum.edu.pk

Abstract. This paper give an insight on effect of education and socio economic factors on education on earning for Pakistan using data mining technique Regression tree and classification tree (CART). Labor force survey data used in this paper. Variables used as predictors in the study are Education, Gender, Status, Training, and Occupation, Location of working, Training, Experience, Age and Type of industry, where monthly income is used as an independent variable. In case of classification income is divided in Quintiles, which is used as a dependent variable for classification variable. Type of industry, education, age and occupation are found significant variables in both classification and regression tree. Regression trees shows that instead of education type of industry is the most important variable and sex and education are the least important variables. Classification tree also shows that Type of industry is the most significant variable which effects the earning of an individual, then age and occupation of an individual come and education is the least important variable where the rest of predictors play no role in earning of an individual.

Keywords: CART · Classification and regression tree · Pruning · Cross validation

1 Introduction

The distribution of the earnings is an important issue for the improving the socio economic condition of any country, especially when income distribution is skewed. To find the cause of difference in earnings of an individual or to find the determinants of earnings of individual whether personal characteristics play important role in effecting the earning of an individual or labor market characteristics. Once the factors effecting the earning of individual are known, then it is easy to improve life in that country. The predictor schooling used in Mincer earning function for Sweden and different cases when it yields misleading information and its assumptions about length of working life. It was found that the decline in rate to schooling from 1068 to 1981 in college

© ICST Institute for Computer Sciences, Social Informatics and Telecommunications Engineering 2017
J. Ferreira and M. Alam (Eds.): Future 5V 2016, LNICST 185, pp. 215–223, 2017.
DOI: 10.1007/978-3-319-51207-5_22

education where return to high school is stable. There estimate suggests that impact of education on length of working life is an important topic for future research. Education has a causal effect on earnings (Bjorklund 2000).

The factors affecting the earnings of an individual and returns to education for Lahore district Pakistan for teaching and non-teaching staff in university, college and school using multiple linear regressions. The factors that significantly contributed to earning of all employees, university employees, college employees, and school employees were age, experience, occupation, gender, working hour, computer literacy, family background, and spouse education. Those who have passed SSC from private institute earn 8.7 more than those who have passed SSC from Government institute. Family background has positive and significant effect on earnings. Teaching staff earn more than non-teaching staff (Afzal 2011).

Earnings functions for industrial works in Punjab, to analyze the difference in earnings of individuals due to gender, marital status, regional location and other socio economic variables using linear single equation least squares regression analysis (Kapoor and Puri 1971). Parents effect the earning of a child potentially through genes and family environment by using variance component model to find the contribution of genetics, family and environments to the variance of the log earnings of white males around 50. The model is estimated through linear additive equation. The contribution of non-common environment is 46% for the log of earnings and 24% for the years of schooling. After making a lot of assumptions, they partition the remaining variance. Using more plausible estimates, the partitioning of the variance of the log of earning suggests 18 to 41% was due to genetics and 8 to 15% to common environment (Taubman 1976).

Decision tree is a flow-chart-like structure which is used for segmenting or stratifying the predictor space in to a number of regions or subsets, to make prediction for a given value, mean and mode of the training data set is used. The set of splitting rules used to segment the predictors space can be summarized in a tree, this approach is referred as Decision tree methods. The performance of tree based method and linear regression can be assess through test error where test error is estimated through cross validation or validation set. If the pictorial presentation of the model is required than we go for tree based methods. CART technique has used a lot in public health and finance but now-a-days used in economics.

We have used CART for finding the determinants of earning because of its interesting features. The purpose of using regression and classification tree (CART) is unlike simple regression its fit the model at each splitting node of the tree, where simple linear regression fit one model for the complete set of data. The Statistical earning function is given as follows, $\operatorname{Ln} y_i = f(s_i, x_i, z_i) + u_i$, $\ln y_i$: is the log of earning, s_i: is schooling, x_i: is experience, z_i: Represents other factors affecting earning such as training of employees, gender or geographical region of individual, age, hours of work, type of industry the employees are working in u_i: is the disturbance term assumed to be normally distributed (Berndt 1991).

The data used in the study is that of Labor Force Survey 2012–13. We have used the information of only employed persons that is affecting earning of an individual i-e age, occupation, training, gender, experience, residence, educational level, marital status, and income. R-Programming have used for Classification and Regression Tree (CART) to estimate the determinants of earning function.

2 Empirical Results and Discussion

All individuals working in cooperative society, individual ownership, partnership and other, female their average log income is 8.325, so we make the prediction of $e^{8.325}$ i.e. 4125.737. Individuals working in cooperative society, individual ownership, partnership and other sectors but are females and having age less than 20 their average log income is 8.605. So we make prediction of $e^{8.606}$ i.e. 5458.885 but those whose age is greater than 20 their average log income is 9.0306. So the prediction is $e^{9.036}$ i.e. 8400. Those who are working in Government, private and public sector and having education below middle and no formal education and specifically working in private and public sector their mean log income is 9.246 so predicted as $e^{9.246}$ i.e. 10363 but working in the sector other than private and public and age is less than 36 their mean log income is 9.496, i.e. $e^{9.496}$. So prediction is 13306, having age greater than 36 their mean log income is $e^{9.825}$, *i.e.* 18490. So we conclude that the government employees earn more than other sectors employees. Females earn less than male. Employees having higher education and experience, earn most. Those females whose age is greater than 20 earn more than those, whose age is less than 20. This is depicted in Fig. 1.

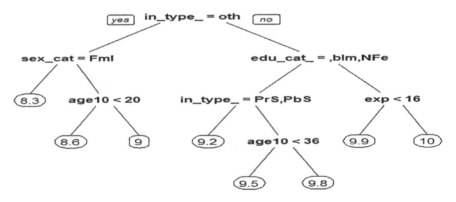

Fig. 1. Regression tree using complete data set of LFS

3 Making Prediction Through Fitted Model Using Testing Data

We have analyzed the data using fitted models as shown in Figs. 2 and 3. Cross validation graph shows the plot of size of the tree against the deviance. We choose that point where deviance approaches to minimum; here the minimum deviance is at size equal to 9. The pruned tree is shorter than the un pruned tree, the important variables are type of industry of an individual, sex and education. Pruned tree has five terminal

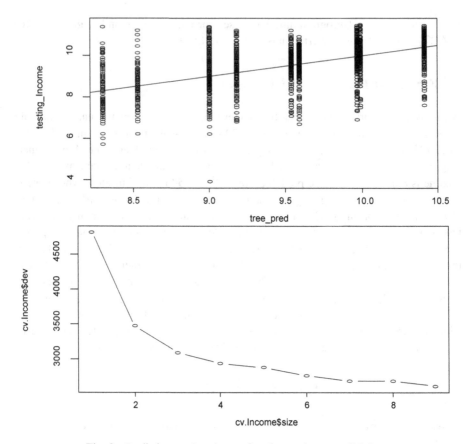

Fig. 2. Prediction made using testing data and cross validation

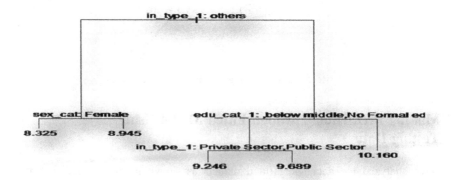

Fig. 3. Plot of prune tree

nodes and three internal nodes. Those individuals who are working in sectors other than government, private and public and are female their average log income is predicted as 8.3 i.e. 4023 and those who are males their average log income is 8.9 i.e. 7331. This shows males earn more than female if they are work in the same type of industry. Those who are working in government, private and public sector and education below middle or no formal education then specifically working in private and public sector their average log income is 9.2 i.e. 9897 and who are working in government or other sector their mean log income is 9.68 i.e. 15994. It shows that government employees with higher education earn more than employees of other sectors. Those who are not working in government, private and public sector in other words working in other sector and education greater than middle their mean log income is 10.160 i.e. 25648.

Figure 4 shows that the prediction using testing data through pruned model is quite good and the numerical measure used for calculating the error of the fitted model is MSE, which in this case is 36% and is increased a little for testing data.

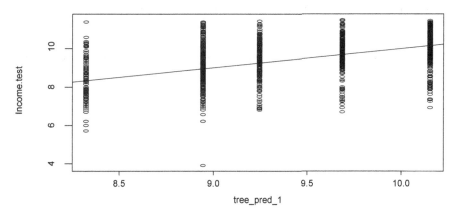

Fig. 4. Prediction through pruned tree using testing data

4 Classification Tree for Quintiles of Income

The classification tree depicted in Fig. 4, shows that those individuals who are working in government, private and public sectors, blue and pink collar workers and then working specifically in government sector are belongs to the Q1, the 1st quintile of income group. The range of 1st quintile is (0–16428), and those who are working in public, private or other sector also belong to Q1, the 1st quintile of income. Those who belongs to occupation category "white collar job" and "other", age is less than 38.5 and education below middle belong to 1ist quintile of income. Those whose age is less than 38.5 but having education above middle and other also fall in 1st quintile of income.

Those whose age is greater than 38.5, education below middle and no formal education fall in 4th quintile (23273–29275) of income and those whose education category is other than below middle and no formal education fall in 5th quintile

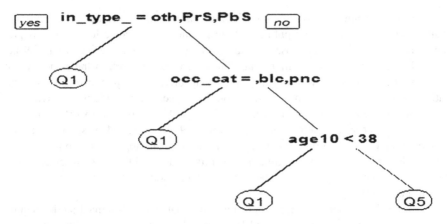

Fig. 5. Classification tree using R-part

(29276–46424) of income as shown in Fig. 5. So we conclude that individuals of "other sector" earn less than government, private and public sectors and belong to lower income group. Those who are working in government, private and public sectors, white collar workers, age greater than 38 and education below and above middle belong to the upper income group. Complete classification tree using training data in shown in Fig. 6.

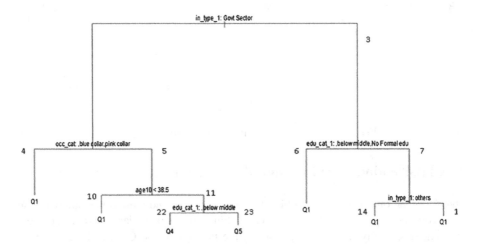

Fig. 6. Classification tree using training data

5 Cross Validation

Plot of the size of the tree against the misclassification shows the different size of tree against the misclassification but best point of pruning is 5, size of tree mean the number of leaves we have or the level to reach in pruning, when size of the tree is 5 the misclassification error is minimum. Figure 7 shows cross validation.

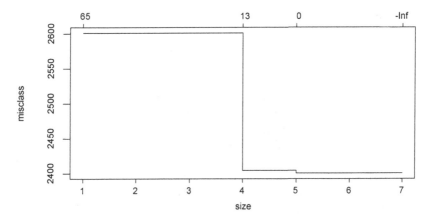

Fig. 7. Cross validation

6 Prediction Using Prune Tree

Left branch of the tree: Those individuals who are working in government sectors, blue and pink collar workers belong to the Q1, the 1st quintile of income group. The range of 1st quintile is (0–16428), and those who are white collar workers and their age is less than 38.5 also belongs to Q1 but those whose age is greater than 38.5 and having education middle or no formal education also belongs to Q4 but those whose education is above middle belongs to Q5 (5th quintile (29276–46424)). Individuals who are working in public, private or other sectors belong to Q1, the 1st quintile of income. Prediction by using pruned model for testing data/unseen data; Error is still 31% so the model is good fit for training and testing data (Fig. 8).

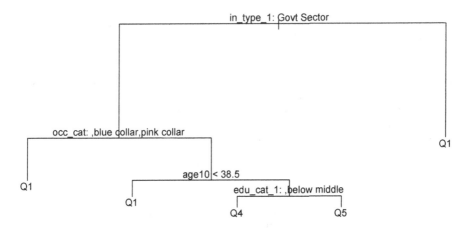

Fig. 8. Prune classification tree

7 Conclusion

In the final prune tree which is free from the problem of over fitting, three variables are significant, type of industry, sex and education. Here age has pruned and show that those individuals who are working in government, private and public sector and having higher education earn more than those whose education is lower than middle and those who are working in government sectors earn more than private and public sector. Female of Individual of cooperative society, individual ownership, partnership and other sectors earn less than male of these sectors. In case of classification, there are Quintiles of income, which is a qualitative variable. The classification tree made for Quintiles predicted that those individuals who are working in government, public and private sectors and are blue and pink collar workers then specifically in government sector, belongs to lower group income and if they are working in private and public sector also belong to lower group income. Those individual who are working in cooperative society, individual ownership, partnership and other sectors belong to lower income group if they are white collar workers and their age is greater than 38, having education above middle they belong to the highest income group but if individual have below middle education or no formal education and have age greater than 38, white collar worker and working in government, private or public sector belongs to income group Q4. So we conclude from classification tree that those individuals whose age is greater than 38, doing white collar job, having higher education and working in Government, private and public sector earn more. Type of industry an employee is working, occupation; education and age are important variable in the study.

References

Bjorklund, A., Kjellstrom, C.: Estimating the return to investments in education: how useful is the standard Mincer education? Econ. Educ. Rev. **21**, 195–210 (2000)

Metcalf, D.: The determinants of earnings changes: a regional analysis for the U.K., 1960–68. Int. Econ. Rev. **12**(2), 273–282 (1971)

Afzal, M.: Micro econometric analysis of private returns to education and determinants of earnings. Pak. Econ. Soc. Rev. **49**(1), 39–68 (2011)

Khan, S., Irfan, M.: Rates of returns to education and the determinants of earnings in Pakistan. Pak. Dev. Rev. **XXIV**(3&4), 671–683 (1985)

Tubman, P.: The determinants of earnings: genetics, family, and other environments; study of white male twins. Am. Econ. Assoc. **66**(5), 858–870 (1976)

Nasir, Z.: Determinants of earnings in Pakistan: findings from the labor force survey 1993–94. Pak. Dev. Rev. **37**(3), 251–274 (1998)

Kapoor, B.L., Puri, A.K.: The determinates of personal earnings: a study of industrial workers in Punjab. Econ. Educ. Rev. (1971)

Sutton, C.D.: Classification and regression trees, bagging, and boosting. In: Hand Book of Statistics, vol. 24 (2005)

Pakgohar, A., Tabrizi, R.S., Khalili, M., Esmaeili, A.: The role of human factor in incidence and severity of road crashes based on the CART and LR regression: a data mining approach. Procedia Comput. Sci. **3**, 764–769 (2010)

Lewis, R.J.: An introduction to classification and regression tree (CART) analysis. In: Annual Meeting of the Society for Academic Emergency Medicine in San Francisco, California (2000)

Berndt, E.R.: The Practice of Econometrics: Classic and Contemporary. Addison-Wesley, Boston (1991)

De'ath, G., Fabricius, K.E.: Classification and regression trees: a powerful yet simple technique for ecological data analysis. Ecology **81**(11), 3178–3192 (2000)

Gordon, L.: Using classification and regression trees (CART) in SAS® enterprise miner TM for applications in public health. In: Data Mining and Text Analytics: 089-2013 (2013)

Horning, N.: Introduction to decision trees and random forests. Am. Mus. Nat. Hist. (2013)

James, G., Witten, D., Hastie, T.: An Introduction to Statistical Learning: With Applications in R. Taylor & Francis, Abingdon (2014)

Liaw, A., Wiener, M.: Classification and regression by randomForest. R News **2**(3), 18–22 (2002)

Loh, W.Y.: Classification and regression trees. Wiley Interdisc. Rev.: Data Min. Knowl. Discov. **1**(1), 14–23 (2011)

Ohno-Machado, L., et al.: Decision trees and fuzzy logic: a comparison of models for the selection of measles vaccination strategies in Brazil. In: Proceedings of the AMIA Symposium. American Medical Informatics Association (2000)

Patel, H.D., et al.: Cost-effectiveness of a new rotavirus vaccination program in Pakistan: a decision tree model. Vaccine **31**(51), 6072–6078 (2013)

Rokach, L.: Data Mining with Decision Trees: Theory and Applications. World scientific, Singapore (2007)

Thakur, G.S., et al.: Understanding the applicability of linear & non-linear models using a case-based study. International Journal of Artificial Intelligence & Applications (IJAIA) **5**, 1–15 (2014)

Varian, H.R.: Big data: new tricks for econometrics. J. Econ. Perspect. **28**, 3–27 (2014)

Chang, Y.: Robustifying Regression and Classification Trees in the Presence of Irrelevant Variables. ProQuest, Ann Arbor (2008)

Friedman, J.H.: Greedy function approximation: a gradient boosting machine. Ann. Stat. 1189–1232 (2001)

Zhang, D.: Advances in Machine Learning Applications in Software Engineering. IGI Global, Hershey (2006)

Author Index

Printed in the United States
By Bookmasters